Synthesis Lectures on Engineers, Technology, & Society

Series Editor
Caroline Baillie, School of Engineering, University of San Diego, San Diego, CA, USA

The mission of this Lecture series is to foster an understanding for engineers and scientists on the inclusive nature of their profession. The creation and proliferation of technologies needs to be inclusive as it has effects on all of humankind, regardless of national boundaries, socio-economic status, gender, race and ethnicity, or creed. The Lectures will combine expertise in sociology, political economics, philosophy of science, history, engineering, engineering education, participatory research, development studies, sustainability, psychotherapy, policy studies, and epistemology. The Lectures will be relevant to all engineers practicing in all parts of the world.

Roger V. Gonzalez

Adapting Engineering Education to a Rapidly Changing World

 Springer

Roger V. Gonzalez ⓘ
Chair and Professor, Engineering Education
and Leadership, College of Engineering
The University of Texas at El Paso
El Paso, TX, USA

ISSN 1933-3633　　　　　　ISSN 1933-3641　(electronic)
Synthesis Lectures on Engineers, Technology, & Society
ISBN 978-3-031-78907-6　　　ISBN 978-3-031-78908-3　(eBook)
https://doi.org/10.1007/978-3-031-78908-3

© The Editor(s) (if applicable) and The Author(s), under exclusive license to Springer
Nature Switzerland AG 2025

This work is subject to copyright. All rights are solely and exclusively licensed by the Publisher, whether the whole or part of the material is concerned, specifically the rights of translation, reprinting, reuse of illustrations, recitation, broadcasting, reproduction on microfilms or in any other physical way, and transmission or information storage and retrieval, electronic adaptation, computer software, or by similar or dissimilar methodology now known or hereafter developed.
The use of general descriptive names, registered names, trademarks, service marks, etc. in this publication does not imply, even in the absence of a specific statement, that such names are exempt from the relevant protective laws and regulations and therefore free for general use.
The publisher, the authors and the editors are safe to assume that the advice and information in this book are believed to be true and accurate at the date of publication. Neither the publisher nor the authors or the editors give a warranty, expressed or implied, with respect to the material contained herein or for any errors or omissions that may have been made. The publisher remains neutral with regard to jurisdictional claims in published maps and institutional affiliations.

This Springer imprint is published by the registered company Springer Nature Switzerland AG
The registered company address is: Gewerbestrasse 11, 6330 Cham, Switzerland

If disposing of this product, please recycle the paper.

To all engineering academics on both sides of the Atlantic who seek to change our system for the betterment of society.

Foreword

As technology advances at an ever increasing rate and problems become more complex, engineering education stands at the crossroads and genesis of change, demanding urgent reform to meet the evolving needs of our global society. This monograph delves into the criticality of such reform by examining the unique challenges faced by educational systems in both the UK and the USA. By exploring the historical context, current landscape, and future directions, the aim is to illuminate a path forward for engineering education.

Historically, engineering education has been the bedrock of technological and industrial advancement. In the UK, the Industrial Revolution heralded the rise of engineering as a profession, with educational institutions evolving to meet the demands of a rapidly changing world. Similarly, in the USA, the post-war era saw a surge in engineering programs designed to fuel innovation and economic growth. While robust, the educational systems supporting engineering have struggled to stay in step with the pace of technological advancements and shifting societal needs of the 21st century.

After earning my degrees from institutions in the USA and spending time in industry, I started my academic career at Texas A&M University where I was tenured and promoted. I then spent 8 years in Wales, UK at Swansea University as a faculty member, department head and, eventually, dean. Currently, I serve as Dean of the College of Engineering at The University of Texas at El Paso. Serving in these roles in both the USA and UK allowed me to become very familiar with the strengths and challenges of higher education systems in both countries. While the engineering education systems in both countries seek to produce engineers with the skills to drive technological innovation, there are significant differences in their structure and approaches. Surveying and assessing the relative strengths and challenges provides not only a current perspective on two of the world's top engineering education systems but also points toward how the systems must evolve to meet societal needs.

Today, the UK and USA face a myriad of distinct yet interconnected challenges in engineering education. In the UK, there is a pressing need to diversify the engineering workforce and make the curriculum more inclusive and reflective of contemporary issues such as sustainability and digitalization. Meanwhile, the USA grapples with issues of accessibility and retention, particularly among underrepresented groups. Both systems must also address the growing demand for interdisciplinary and professional skills as well as the integration of emerging technologies into the curriculum.

Looking to the future, engineering education reform must be holistic and forward-thinking. It requires a concerted effort to bridge the gap between academia and industry while ensuring that graduates are technically proficient and equipped with the professional skills necessary for leadership and innovation. Furthermore, there must be a global perspective that recognizes engineering challenges and solutions that transcend national borders.

This monograph seeks to provide a comprehensive overview of these issues with particular focus on the UK and USA, offering insights and recommendations for engineering educators, policymakers, and industry leaders. By reflecting on past and current societal contexts, critically analyzing the present, and envisioning the future, I believe the monograph can meaningfully contribute to the ongoing dialogue on engineering education reform and inspire substantive change.

Kenith E. Meissner II, Ph.D., FInstP
Dean and Riter Professor of Engineering
College of Engineering
The University of Texas at El Paso
El Paso, TX, USA

Preface

For almost 30 years, engineering education has been one of the focal points of my professional academic career. I spent 16 years at LeTourneau University—a small, private college known for its close-knit community and hands-on approach to engineering education—to my current position of 13 years at The University of Texas at El Paso (UTEP), a large public institution recognized as the only open access and Carnegie Tier 1 research university in the USA. Given that these two universities could not be more different, I have experienced firsthand the diversity of educational environments in the USA. This journey has been enlightening and challenging, as each institution brings unique strengths, opportunities, and obstacles to overcome. The richness of these experiences has deepened my appreciation for the complexities of engineering education and sparked a deep curiosity to explore it more broadly, especially in comparison with the systems of other countries, particularly in the UK, with an emphasis on England.

The history of engineering education is long and layered, stretching back centuries and evolving through industrial revolutions, wars, and waves of technological advancement. Many of the foundational practices in U.S. engineering education were inherited from the UK, which pioneered the structured training and apprenticeship models that laid the groundwork for modern engineering as a profession. Over time, these practices were adapted to suit the American context, blending elements from a liberal arts tradition with the technical rigor required by industry. Yet, despite this shared heritage, significant differences have emerged between our systems, shaped by cultural values, economic priorities, and educational philosophies. These differences have long fascinated me. What are the fundamental distinctions that set our systems apart? How do these distinctions impact how we teach, the skills we emphasize, and the expectations we set for our students? And most importantly, how do they shape the professional identities and capabilities of the engineers we graduate?

A core question I sought to answer through this work was how our respective US and UK systems approach the promotion of professional skills within undergraduate engineering education. Professional skills—often referred to as "soft skills"—are communication, teamwork, leadership, ethics, and adaptability, among many other skills complementing technical competence. These skills are essential to engineers who must collaborate across disciplines, lead teams, and consider the societal implications of their designs. Yet the emphasis and approach to teaching these professional skills can vary widely. In some cases, they are woven into the curriculum through experiential learning, while in others, they are assumed to develop naturally alongside technical skills. My goal was to understand the nuances of how each system approaches this critical aspect of engineering education and what we might learn from one another to better prepare our students for the demands of modern engineering careers.

Within the USA, our higher education landscape is incredibly diverse, encompassing a wide range of institutions, from large research universities to small liberal arts colleges, community colleges, and technical institutes. Each type of institution plays a unique role in educating engineers, and this diversity provides students with numerous pathways into the profession. However, this also makes it challenging to generalize "the American engineering education system." The UK's system, particularly in England, has its own distinct structure and types of institutions, each with a unique history and purpose. Understanding these differences—both within our own country and across the Atlantic—is essential if we are to have a meaningful comparison. This monograph reflects my year-long efforts to gain a deeper understanding of each system's structural and philosophical differences, especially as they pertain to preparing engineers for a globalized world.

As an educator, I am frequently asked, "Which system is better?" It's a natural question, yet the answer is far from straightforward. There is no simple metric for comparing such complex systems; each is shaped by a unique interplay of history, government policies, private accreditation bodies, and economic demands. In many respects, one system's strengths mirror the other's challenges. For example, while the US system benefits from flexibility and variety, it also contends with issues of uneven quality and accessibility. Meanwhile, the UK's system offers a more standardized approach, which can streamline specific processes but may also limit institutional innovation. Ultimately, I believe that both systems have inherent strengths and are tuned to their respective societies and institutional politics. Still, both also face significant challenges as we move into a future that demands adaptability and resilience from our engineers.

As engineers, we are trained to be cautious and methodical, knowing that our work directly impacts human welfare and safety. This caution often extends to our approach to change within the profession and within our educational systems. When changes do occur, they are often incremental, introduced only after careful consideration and rigorous testing. However, we live in an era where technological advancements and societal shifts are happening at an unprecedented pace. In this ultra-rapid change environment, engineering education risks being left behind, struggling to keep up with the evolving demands of industry and society. If we are to prepare the next generation of engineers to thrive in this world, we must be willing to examine our history and our current practices critically.

By understanding the roots of our educational systems, we can make informed decisions that allow us to innovate responsibly without repeating the mistakes of the past.

This monograph is intended as a resource for educators, administrators, and policymakers who are passionate about advancing engineering education. It is my hope that by exploring the differences and similarities between our systems, we can foster a more collaborative dialogue between the US and the UK, one that acknowledges the strengths of each approach while addressing the challenges that lie ahead. Engineering is, at its heart, a global profession. The problems our graduates will face—climate change, energy sustainability, infrastructure resilience—are not confined by borders. To equip them for this world, we must draw from the best practices of each system, learning from one another and working together to create an education that is as adaptable and innovative as the engineers we aim to cultivate.

May this monograph serve as a stepping-stone toward a deeper understanding and a more unified vision for engineering education across borders. Whether you are an experienced educator, a newcomer to the field, or someone with a vested interest in the future of engineering, I hope that these reflections and analyses will provide insight, spark ideas, and contribute to the ongoing evolution of our shared profession.

Roger V. Gonzalez, Ph.D., PE, F.ASME,
F.AIMBE
Chair and Professor
Engineering Education and Leadership
College of Engineering
The University of Texas at El Paso
El Paso, TX, USA

Acknowledgments Deep appreciation goes to Prof. John Mitchell at the University of College London (UCL) and Prof. Roger Penlington at Northumbria University for their guidance before, during, and after this project. Their insights were extremely valuable. This project was undertaken under the UT System Faculty Fellowship program with significant funding from the Bob and Diane Malone Gift Fund and partially through the Centre for Engineering Education, UCL Engineering. Deep gratitude to Dr. Annie Olson for her expert review and valuable feedback.

Contents

Engineering Education in the UK and the US: A Comparative Analysis.
The Complexity of Modern Engineering 1
    Engineering Education in a Global Context 3
    Research Goals and Analysis Approach 4
    Research Findings 6
        Programs 9
        Students 20
        Faculty 25
        External Factors 29
    Conclusion 35
    References 35

Preparing Engineering Students for Leadership 43
    Engineers as Leaders 43
    Professional Development 45
    Broadening Access and Participation in Engineering Education 61
    What Comes Next? 68
    References 69

Engineering Education in the Age of Accelerations 75
    Acceleration of Technology 76
    Acceleration of Globalization 77
    Acceleration of Environmental Changes 78
    Engineering Education at the Forefront of Change 79
    Artificial Intelligence in Engineering Education 81
    Four Functions Higher Education Must Continue to Fulfill 83
    Validation/Certification 83
    Structure and Process of Education 84
    Mentorship 84
    Community 85
    References 86

About the Author

Roger V. Gonzalez is the inaugural chairman and professor of the Department of Engineering Education and Leadership at The University of Texas at El Paso (UTEP). He is the Mike Loya Endowed Chair in Engineering. He also serves as the Director of the Engineering Leadership program for the College of Engineering. Dr. Gonzalez earned a B.S. in Mechanical Engineering in 1986 from UTEP. He earned his M.S. in Biomedical Engineering and Ph.D. in Mechanical Engineering from The University of Texas at Austin. He was a Post-Doctoral Fellow at the premier Rehabilitation Institute of Chicago and Northwestern Medical School. Professor Gonzalez has been recognized for scholarly work, education innovation, and socio-entrepreneurial humanitarian efforts. He is known and respected for actively incorporating students into these three areas.

Among many highlights of his scholarly work, he is a Fellow of the American Society of Mechanical Engineers (ASME) and a Fellow of the American Institute for Medical and Biological Engineering (AIMBE) and was awarded a prestigious National Institutes of Health (NIH) National Research Service Award for his work in neuromuscular control and musculoskeletal biomechanics in children with juvenile rheumatoid arthritis. Dr. Gonzalez's scholarly work includes over 100 publications in journals and conference proceedings.

For his efforts and innovation in engineering education, Dr. Gonzalez has received the American Society of Engineering Educators (ASEE) Teaching Award, the Minnie Stevens

Piper Foundation Award, and LeTourneau University's top research and scholarship award. He was also a Finalist for the IEEE Global Humanitarian Engineer of the Year award in 2013. He serves as an engineering program evaluator for the Accrediting Board for Engineering and Technology (ABET).

Dr. Gonzalez was awarded a faculty fellowship by UTEP and the University College London (UCL), where he served as a Visiting Professor, to spend the 22–23 academic year traveling throughout the UK visiting over 25 universities in England, Wales, Scotland, Ireland, and The Netherlands. His research focused on how professional development plays a role in engineering education and how approaches to broadening access impact higher education.

Dr. Gonzalez is the Founder, former CEO and President of LIMBS International (www.limbs.org), one of the most recognized international providers of low-cost prosthetic components. LIMBS is a 501(c)3 non-profit humanitarian organization that designs, creates, and deploys prosthetic devices to transform the lives of amputees in the developing world by restoring their ability to walk. Since its founding in 2004, the LIMBS Knee has helped thousands of amputees in over 50 countries on four continents.

Author Details:
Roger V. Gonzalez, Ph.D., P.E., F.ASME, F.AIMBE
The University of Texas at El Paso
Mike Loya Endowed Chair in Engineering
Inaugural Chairman and Professor, Department of Engineering Education and Leadership
Inaugural Director—Bachelor of Science in Engineering Innovation and Leadership
Engineering Disciplines: Mechanical and Biomedical Engineering
Founder and former CEO/President—LIMBS International

Engineering Education in the UK and the US: A Comparative Analysis. The Complexity of Modern Engineering

Engineering provides solutions that are imperative to human well-being and to societal safety, growth, and prosperity. Engineers provide solutions to known as well as potential problems. However, despite engineering's extraordinary successes, engineers still learn from catastrophic failures.

On October 17, 2000, a train from London bound for Leeds derailed near Hatfield. Four people lost their lives, and over 70 were injured due to "a series of errors by rail bosses and engineers" [68]. Gauge corner cracking (GCC) led to a rail breaking and causing the crash. The problem with the rail had been identified nearly two years earlier; the replacement rail was sitting beside the track, waiting to be installed when the crash occurred [11]. This was not a technical error, but an engineering management and execution error that cost human lives.

Engineering firm Balfour Beatty and rail operator Network Rail were charged with a health and safety violation and four counts of manslaughter [11]. The manslaughter charges were later dismissed, but the judge imposed fines totaling 18.8 million GPB (24.3M USD) on Balfour Beatty and Railtrack—5 times the previous record in England—calling the crash "the worst example of sustained negligence in a high-risk industry I have ever seen" [34].

Another failure occurred in the US in May 2019. The US Army Corps of Engineers released water from the Keystone Dam at a rate more than double the flow of Niagara Falls. In the already rain-soaked areas below the dam, water breached the levees on three rivers, causing severe flooding in communities across two states. More than 500 homes were damaged or destroyed. Crops were ruined just two weeks before harvest. Businesses and industries suffered staggering losses.

Residents whose homes were ruined argued that the Corps should have anticipated the spring rains and slowly lowered water levels months earlier. A representative for the Corps disagreed, claiming that a precautionary release could have wreaked havoc if rain fell simultaneously downstream. A Corps official said, "The systems worked the way they were designed, the way they were engineered. I think our people made really good decisions" [28]. This event happened not because of technical miscalculation but due to the lack of a broader understanding of the environment by engineers.

An article listing "the worst engineering disasters" insightfully sums up the crucial contributions and responsibilities of engineers:

> The engineering field has contributed immensely toward changing how the world works. In many ways, the innovations and inventions of the last few decades have been nothing short of incredible. However, there have also been some tragic and unforgettable **engineering catastrophes** [emphasis in original]. These disasters have generally resulted from a mixture of design failures, under or overestimations, acting on _insufficient knowledge_, and _other factors_ [emphasis added]. Nevertheless, these disasters are also an opportunity to learn from our mistakes so as not to repeat them in the future. [85]

A root cause analysis of another engineering disaster, the 1981 Hyatt Regency Hotel collapse, explains that "to function in the modern world, one must often place trust in engineers" [64]. Contemporary engineers make decisions extending far beyond designing, manufacturing, and building. Their technical expertise must be augmented by a broad understanding of health and safety protocols, public policy, business management, law, environmental concerns, sustainability, weather patterns, communication, public policy and politics, economics, globalization, ethics, and much more.

The complexity of modern engineering is not unexpected. In 2004, members of the US National Academy of Engineering collaborated to publish *The engineer of 2020: Visions of engineering in the new century*. They argued that engineering education and the education profession cannot remain static:

> Society continually changes and engineering must adapt to remain relevant. But we must ask if it serves the nation well to permit the engineering profession and engineering education to lag technology and society, especially as technological change occurs at a faster and faster pace. Rather, should the engineering profession anticipate needed advances and prepare for a future where it will provide more benefit to humankind? Likewise, should engineering education evolve to do the same? (National Academy of Engineering [55, p. 1])

Obviously, the answer to both questions is "Yes." But an even more important question remains: How?

Engineering Education in a Global Context

After 37 years as an innovator and entrepreneur, engineer, and engineering educator, I am now the Inaugural Director, Inaugural Chair, and Professor in a new Bachelor of Science in Engineering Innovation and Leadership (E-LEAD) program at The University of Texas El Paso (UTEP). In 2012, as my colleagues and I designed the E-LEAD program, the first of its kind in the US, we consulted with numerous engineers in exploring questions of how engineering education must adapt to our rapidly changing world—not only technologically but also in widening access to engineering programs and broadening the capabilities of engineering graduates—understanding that while technological know-how is still necessary, *it is increasingly insufficient*. Over the last decade, through my journey in designing and implementing our E-LEAD program and its educational objectives, I learned from programs in the US that aimed to change the paradigm of engineering education at a foundational and innovative level. For example, Olin College collaborated with us in the early stages of our program design.[1]

Given my knowledge and experience within the US system and having traveled the world enough to know that the philosophy of higher education varies across the globe, I wanted to learn how engineering programs around the world were responding to global challenges, especially within the United Kingdom (UK).

Why the UK? The United Kingdom and the United States have a long and enduring partnership based on shared values and interests, including a commitment to academic excellence in higher education. Both have a long history of providing quality higher education, and both the UK and the US are world leaders in higher education. Both countries' universities attract students worldwide and produce some of the most innovative and groundbreaking research. Also, our shared English language facilitates dialogue with academics and staff. Despite these many similarities, the two systems have some critical differences. The fact that we are similar in our historical educational legacy yet

---

[1] For more information on the development of the E-Lead program, see the following:

R. T. Schoephoerster and P. Golding, "A New Program in Leadership Engineering," *2010 IEEE Transforming Engineering Education: Creating Interdisciplinary Skills for Complex Global Environments*, Dublin, Ireland, 2010, pp. 1–17, https://doi.org/10.1109/TEE.2010.5508824.

R.V. Gonzales et al., "Model Collaboration for Advancing Student-Centered Engineering Education," *2013 IEEE Frontiers in Education Conference (FIE)*, Oklahoma City, OK, USA, 2013, pp. 212–214, https://doi.org/10.1109/FIE.2013.6684819.

R.V. Gonzalez et al., "The Creation and Inauguration of Engineering Leadership: UTEP and Olin College innovation project," *2015 IEEE Frontiers in Education Conference (FIE)*, El Paso, TX, USA, 2015, pp. 1–8, https://doi.org/10.1109/FIE.2015.7344095.

Golding, P. et al., (2021, July), *Leadership in Engineering Innovation and Entrepreneurship* Paper presented at 2021 ASEE Virtual Annual Conference Content Access, Virtual Conference. https://peer.asee.org/37422.

still significantly different provides the opportunity for a comparative study, identifying strengths and weaknesses that will enable enhancements to both systems.[2]

Towards that end, I consulted with the University College London's (UCL) Center for Engineering Education (CEE), Dr. John Mitchell, Professor of Communications Systems Engineering and Co-Director of the UCL-CEE, and with the leadership of the Engineering Research Network (ERN), Drs. Roger Pennington and Claire Lucas. Those discussions led to framing a year-long faculty research fellowship to study how engineering education impacts and is impacted by broader global contexts.

Research Goals and Analysis Approach

I wanted to learn how engineering programs in the UK, primarily in England and Wales, are approaching innovative aspects of engineering, especially by examining how universities in the UK are (1) dealing with students' professional development and (2) widening the access and participation of underserved populations. Professional development is essential training, preparing engineers to work with other professionals across disciplines to contribute to the kinds of complex solutions engineers are called upon to provide. In our E-LEADprogram at UTEP, we identify the following ten skills we want to develop in our graduates:

1. *Multi-disciplinary engineering*: Apply knowledge and skills related to foundational engineering concepts, designs, and analysis that align with students' interests and passions (including passions outside of engineering).
2. *Exemplary leadership*: Enacted by five key principles—Model the way, inspire a shared vision, challenge the process, enable others to act, and encourage the process.
3. *Innovation and entrepreneurship*: Identify, design, and deploy innovative socio-technical solutions that address issues of desirability, feasibility, and viability.
4. *Business acumen*: Employ an understanding and ability to deal with business scenarios in a manner that leads to successful outcomes, with knowledge and experience in foundational business principles of accounting, economics, finance, management, and marketing.
5. *Teamwork*: Contribute as an effective, influential group member to meet a common goal.
6. *Communication*: Reach mutual understanding through effectively exchanging information, ideas and feelings.
7. *Comfort with ambiguity*: Learning that both life and projects can come with significant ambiguity, and that adapting to operate within it is crucial.

---

[2] Although part of the UK, Scotland has an educational system that is more similar to the system in the US.

8. *Adaptability*: Readily adjust to changing and complex situations, acquiring necessary skills and knowledge along the way.
9. *Critical thinking*: Analyze and evaluate issues to understand problems and develop innovative solutions.
10. *Problem-solving*: Find innovative solutions to difficult or complex issues.

These professional development goals identify the knowledge and skills we believe are necessary for engineers—beyond their subject-area expertise—to contribute meaningfully to developing the solutions needed by our rapidly-changing world.

The second key focus, widening access and participation of underserved populations, describes the goal to increase opportunity and achieve equity for students from different backgrounds, underserved people groups, and constrained economic environments. Both the UK and the US are experiencing shortfalls of qualified engineers. A 2021 Skills Survey by the Institution of Engineering and Technology (IET) found that in the UK, "In engineering specifically … two-thirds of those questioned reported gaps they were having trouble filling" [23]. In the US, a report based on data compiled by the US Bureau of Labor Statistics states that between 2016 and 2024 America will be over six million short of the number of engineers needed" [31]. Rutgers University School of Engineering Interim Dean Alberto Cuitiño states that in the US, "we are simply not graduating enough engineers to meet the demand" (as cited in Daks [20]). Broadening access and participation is a critical component of recruiting and training the engineers needed in both the UK and the US.

My intention is to provide a fruitful context for UK and US engineering educators to analyze and question the reasoning behind our practices and to review and share best practices so we can provide the best possible educational experience for our students, especially in terms of professional development and broadening access and participation.

This book does not follow the traditional pattern of quantitative, data-driven engineering scholarship. Instead, it presents findings and broad conclusions based on the summary of qualitative evidence from over 200 interviews and several months of observations. These findings are also supplemented and supported by research in particular areas. These general findings are significant because they provide a broad overview of the stated topics and can be used to inform policy decisions or to develop new research questions. However, it is important to remember that general findings are not always applicable to specific populations or situations. I have no doubt that an exception could be found for every observation; in fact, I encountered some exceptions, but the general observations were still clearly apparent.

Research Findings

The findings presented here are based on visiting almost 30 academic universities with one to three (in a few cases more) visits to each institution selected from various universities, primarily in the Southern, Midlands, and Northern regions of the UK (see Fig. 1). The institutions selected for study were based partly on the guidance of UK academic colleagues, while others were chosen at my discretion. All visits were subject to the availability of academic colleagues at such universities to host and coordinate my visit(s) with various academics at their institution. Over 90% of my visits were onsite at the institution. (Some follow-ups were conducted via video conferencing.) My conclusions are minimally applicable to institutions in the Netherlands and Scotland because the US system has stronger links to the Scottish system. Also, I will not name the universities visited or the academic/administrative personnel with whom I had discussions per the agreement that all conversations were strictly confidential to gain a more candid and unrestricted view of their own academic colleagues and environment, their policies and subsequent impact, and the overall state of engineering education in the UK.

Figure 2 illustrates the group/classification of colleges and universities visited. Schools in the *Russell Group* made up 35% of those visited. The Russell Group is the most comparable to the Ivy League in the US with one notable exception: universities may join or leave the group. However, they rarely do so. The membership, made up of 24 UK universities, has not changed since 2012. The Russell Group is self-selecting; universities can become members based on the quality of research, graduate employment, student satisfaction and other factors. The Russell Group exerts a powerful influence on education policy in the UK. It opposes cuts to government funding for higher education and promotes increased investment in research. Russell Group universities place a high value on internationalization; they attract and recruit researchers and students worldwide.

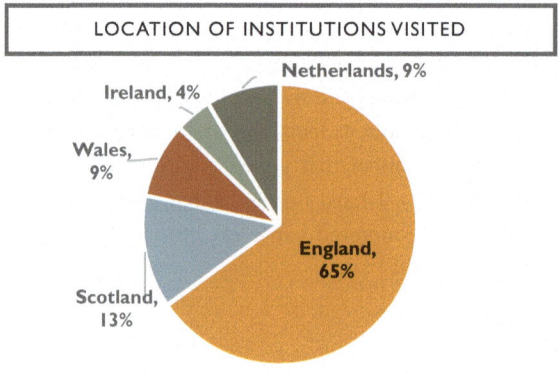

Fig. 1 University locations

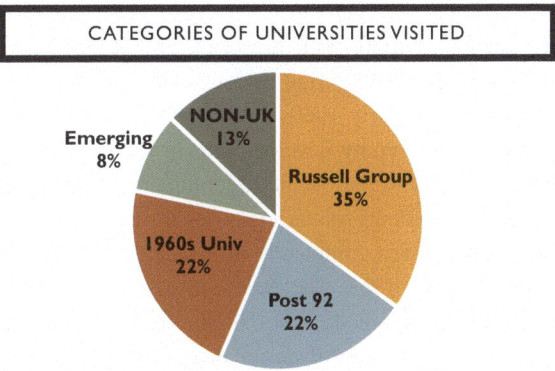

Fig. 2 University categories

The small system of private universities in the UK exerts minimal influence. In the US, the private university system is highly influential. This is especially true of the Ivy League, comprised of eight private research universities. The term *Ivy League* dates back to 1933, but it became official in 1954 as the name of a collegiate athletic conference. All but one (Cornell University) of the members were founded as colonial colleges before the American Revolution. The term Ivy League now describes a group of universities known for highly selective admissions and excellence in academics and research. Membership in the Ivy League is fixed at eight universities. (Some lists cite twelve members, but only because many people mistakenly think that the University of Chicago, Duke University, Stanford University, and the Massachusetts Institute of Technology are Ivy League schools. They are not.) All eight Ivy League schools are also members of the prestigious Association of American Universities (AAU). The AAU strongly supports the quality of education and research. Membership is by invitation only. AAU schools, along with schools in the Russell Group, are considered among the most prestigious universities in the world.

The *Post 92* category includes institutions that became universities after the passage of the Further and Higher Education Act 1992. This act allowed thirty-five polytechnics to become universities and to award their own degrees. The change affected more significant factors than the name of the institution, but in this comparison of UK and US universities, the name change is worthy of note. In the UK, the word *university* carries more prestige than the word *college*, denoting higher rather than further education, but in the US, the words are often used interchangeably. In fact, the Forbes list of "America's Top Colleges" includes mostly universities, including those in the Ivy League [90]. Many American institutions named as *colleges* award both undergraduate and graduate degrees.

The *1960s Universities* are also known as the plate glass universities because their architectural design contrasts sharply with the brick and stone construction of institutions founded in the medieval and Victorian periods. This group began with universities established in the late 1950s and early 1960s based on the work of the University Grants Committee to plan and fund university development. In 1963, the Committee on Higher Education released the Robbins Report, which recommended that colleges of advanced technology (CAT) be given university status. Along with ten CATs that became universities, between 1961 and 1965, seven completely new universities were formed in the UK. Two more followed in 1967 and 1968. The New University of Ulster, founded in 1968, was the final new university established in the UK through public funding.

Emerging Universities is a grouping of newer institutions that do not fit well into the previous categories. These schools offer degrees, sometimes exclusively, in engineering, but their methodology differs significantly from that of institutions in the other groups.

These categories are useful to identify, in general terms, the types of institutions I visited. As universities in the UK and the US consider when and how to anticipate and respond to rapid global change, these categories may provide additional insight. For now, though, I want to establish a context for the topics that will be addressed in this series' second article, in which we will explore two key issues: students' professional development and widening the access and participation of underserved populations. This first paper establishes that context by comparing and contrasting specific topics in four components of engineering education in the UK and the US: Programs, Students, Faculty, and External Factors.

The UK and the US have two of the most respected engineering educational frameworks in the Western world. Many international students and academics seek to become part of their ethos. While some engineering programs around the world are quite similar to the US model, the Bologna Process has been instrumental in establishing a European ethos of quality and portability through unified educational goals and reforms for European member nations. The Bologna Process began in 1998–1999 "to enhance the quality and recognition of European higher education systems and to improve the conditions for exchange and collaboration within Europe, as well as internationally." Specific goals include standardizing "the three-cycle degree structure (bachelor, master's, doctorate) and adopt[ing] shared instruments, such as the European Credits Transfer and Accumulation System (ECTS) and the Standards and Guidelines for Quality Assurance in the European Higher Education Area (ESG)." All member nations agreed to continue the Bologna Process when the European Higher Education Area was established in 2010. Currently, 49 nations are members, and "the Bologna Process has grown into a Europe-wide policy platform for coordinated higher education reform. It addresses new topics, such as fundamental values and learning and teaching; as well as its longstanding commitments, which require continued attention" [7].

While the Bologna Process establishes overarching standards for outcomes, assessment, and other aspects of European higher education programs, each institution retains its unique qualities. A case study sponsored by the Royal Academy of Engineering noted some important ways that universities differ:

> Relevant factors include geographical location, research or teaching led, vocational or academic focus, nature of the student cohort and the history of/mission for industrial engagement. As a result, each university has its own identity, attracting students from different backgrounds (in terms of qualifications, age, part-time/full-time, work-based and socio-economic background). Each university and engineering department also seeks to produce graduates with a particular mix of skills, knowledge and experience to suit specific areas of the graduate jobs market. (Arlett et al. [3, p. 19])

The unique identity of universities in the UK grows out of a long history that predates both the Bologna Process and the US education system by several centuries. Oxford University dates back to 1096, over five centuries before the first British settlers landed in what is now the US. Harvard University, the first American institution of higher learning, dates its founding to 1636—nearly five and a half centuries after Oxford. Even the oldest American universities are relatively new by comparison. Given its long history of higher education, it is no surprise that the UK has a distinct model with variations between England, Wales, Scotland, and Northern Ireland.

To avoid confusion, clarification of some terminology differences between the UK and the US is needed (see Table 1).

Programs

Program Length
Table 2 describes—in very general terms—differences in the length of programs between the UK and the US [82].

Students who search online will find multiple sites advising them that earning a bachelor's degree is faster and less expensive in the UK than in the US. Similarly, a master's or doctoral degree can generally be completed more quickly in the UK than in the US. Of course, student progress varies, but the funding model in the UK allows less variability. Many US students work and attend school part-time. The flexibility in the US funding model makes this possible while funding in the UK precludes part time studies for most students.

Postgraduate Studies
Engineering students accrue multiple benefits from earning a postgraduate degree. A master's degree "usually serves as a pre-requisite for work as a chartered engineer" [5]. Some students opt for the MEng, but since no bachelor's degree is awarded, it is not equivalent to a master's degree in engineering. The US requires a bachelor's degree either prior to or in conjunction with a master's in engineering.

Table 1 UK and US terminology

	UK	US
A program of study leading to a specific degree	Course	Major
A single class	Module	Class or course
University time units in a year	3 terms (11 weeks each, with little significant teaching in summer terms)	2 semesters (14–15 weeks each, often with additional summer or winter sessions)
A university department	Faculty[a]	Department
A group of university departments	Faculty or College	Various, including division, School, or College
Teachers/professors in a university	Academics	Faculty
Ranks of teachers in a university	• Professor—Usually reserved for the most distinguished, senior academics • Senior Lecturer • Lecturer • Reader[a]	• Full professor[3] • Associate Professor • Assistant Professor • Adjunct Professor

[a]In the UK, a current trend is to move closer to the US model for titles and definitions
In the discussion that follows, terms will be used as appropriate to the context, either the UK or the US

Table 2 Program length

	UK	US
Bachelor's degree	3 years (Longer if a foundation year, sandwich year, or internship year is included)	4 years (Longer at some schools and for some majors, including engineering)
Master's degree	1 year	1.5–2 years
Doctoral degree	3–4 years	4–6 years[4]

---

[3] These ranks typically apply to tenure-track appointments. In an attempt to validate and clarify different roles within a university, institutions are increasingly adopting non-tenure-track rankings such as Instructional Professor, Professor of the Practice, Research Professor, and Teaching Professor.

[4] This longer time frame is associated with a student's moving from a BS directly into a PhD without earning a separate master's degree. This option is becoming increasingly popular in the US.

Table 3 Comparison of employment data for working age non-graduates, graduates and postgraduates in the UK (2022) (with the percentage of change over the previous year)

	Non-graduates	Graduates	Postgraduates
Employment rate	69.6% (− 0.2%)	87.3% (+ 0.6%)	89.3% (+ 1.1%)
High-skilled employment rate	23.6% (+ 0.2%)	66.3% (+ 1.1%)	78.3% (+ 1.1%)
Median nominal salary	£27,000	£38,500	£45,000

Data from the UK government report [32], comparing employment rates and median salaries for non-graduates, graduates, and postgraduates, clearly establishes the value UK employers place on postgraduate degrees (see Table 3).

In the UK, engineering has been the leader in developing the integrated master's degree, which combines the bachelor's and master's degrees into a single program, usually lasting four years (with another year possible for industry experience). Students earn both a bachelor's and a master's degree in the integrated master's degree program. A 2017 article from *The Guardian* states that "engineering departments are leading the way in boosting students' career prospects by offering them the longer course instead of the traditional three-year degree. More than half of integrated master's students are engineers or scientists" [91]. *Masters Compare* agrees, stating that "integrated masters programmes started because there was a lack of qualified engineers" [1].

The integrated master's degree has become very popular with students in the UK. The majority of the institutions I visited offer an integrated master's degree program in engineering. Funding from the undergraduate student finance system offers tuition fees and maintenance loans to cover all four years of study [5].

Combined BS/Master's degree programs are also offered in the US, though they are less popular than in the UK. The benefits of earning a postgraduate degree in the US are similar to those of the UK. Table 4 describes some of those benefits.

Program Design: Focused Versus Broad
Many US educators have difficulty imagining how bachelor's and master's degrees can be completed in the UK in fewer years than the time required for completion in the US. A comparison of UK and US educational programs illustrates that while the UK educational program design focuses specifically on the student's specialization, the US program is designed to include broader learning across disciplines. As Table 5 illustrates,

Table 4 US Engineers: degree attainment and median annual income [25]

	Bachelor's degree	Master's degree	Doctorate
Percentage of US Engineers	65%	13%	[No data]
Median Annual Income	$91,031	$104,300	$113,933

Table 5 A general comparison of UK and US Pre-University educational programs

UK		US	
Primary education	Year R (Reception, ages 4–5) Years 1–6	Primary (or Elementary) Education	Kindergarten through Grade 5
Secondary school	Years 7–8	Middle School (or Junior High) Education	Grades 6–8
Secondary school	Year 9 Transition year from Junior to Senior School	Secondary (or High School) Education	Grades 9–12 May include college-preparatory curriculum or vocational skills training
Secondary education	Years 10–11		
Students earn a General Certificate of Secondary Education (GCSE)		Students earn a High School Diploma	
Age of completion	Age 15 or 16, depending on date of birth	Age of graduation	Age ranges from 17 to 19; the average is 18 years old
A-Level study	Years 12–13 Two years of study to prepare for university admission, culminating in A- (Advanced) level exams	Students choose to work or attend a vocational school or two-year community college or proceed to a 4-year university	

that program design difference shows up even in pre-university education. (The names used to describe educational levels in the US vary among schools and regions, as do the grades included in each level. Table 5 presents a general summary).

Students in the UK and the US study the same broad range of general education topics, including math, science, history, geography, English composition, literature, and possibly a foreign language. They also spend the same number of years—13 (including reception year or kindergarten)—in pre-university education, but there are significant differences.

Students in the UK complete their general education studies and choose a university field of study—essentially their career direction—by the end of year 11 at the age of 15 or 16. A-level studies (years 12 and 13) focus on three or four subjects central to the student's university course, preparing the student to qualify for admission. Because of university admission policies, changing that course after year 11 can be quite difficult and costly for students.

In the US, students continue their general education studies throughout year 13 and even into their university studies. General education courses usually make up about one-third of the requirements for a bachelor's degree. (Some courses overlap with major requirements, possibly reducing this to one-fourth.) Many, if not most, of the classes students take during their first two years (especially the first year) are general education classes. Even at the PhD level, US students have coursework requirements in subjects not explicitly related to their research topic. For example, my PhD in biomedical and mechanical engineering included courses in subjects like bio-rheology, neurophysiology, and laser-tissue interaction—despite my research area being musculoskeletal biomechanics.

In one area, pre-university education in the US is evolving to become more like the UK model: students being urged toward an academic or career direction at an earlier age. Schools are encouraged to offer Career and Technical Education (CTE) classes beginning in the 9th grade and sometimes even as early as middle school. A 2019 US Department of Education report reveals that 77% of high school students earned at least one CTE credit. However, only 37% of those students earned a second CTE credit in the same "career cluster" [8]. Only about 28% of US students complete year 13 with two classes in a specific career-focused subject. By contrast, students in the UK spend two years pre-university studying only subjects intrinsic to their university course.

CTE classes are growing in popularity. The US government offers funding to encourage CTE, and many schools have begun to require students to take at least one CTE course. Yet CTE remains somewhat controversial in the US. Detractors argue that schools are pushing students toward careers too early. Advocates claim that CTE allows students to explore—and reject as well as choose—possibilities for their academic and career directions [29].

Both focused and broad educational strategies have advantages and disadvantages. One advantage of the focus in the UK is that it enables students to complete their university courses more quickly. Depending on contact hours and curriculum, students may also acquire more advanced knowledge and expertise in their specific subject area. This is evidenced by the fact that undergraduate students in the UK often take advanced classes that may not be available to students in the US until graduate school. The UK model encourages students to develop a sense of direction earlier in their academic careers, which may also contribute to professional development goals. As counselor and former academic sociologist Rose [65] explains, "Having direction allows you to maintain mental resilience during transitions by facilitating a sense of underlying purpose, not dependent on the specific role one occupies. Also, having a sense of direction promotes better mental health and stronger adherence to long-term goals." These advantages support a focused model.

The broader educational model also has advantages. US educators question the value of ending general education in year 11 when students are just 15 or 16 years old. When I discussed this with academics in the UK, they generally expressed the belief that students had acquired the necessary breadth of education by the time they completed secondary school. However, US educators tend to agree with the view expressed by former Proctor & Gamble Chairman A. G. [43] that a thorough liberal education is the best preparation for success in any career:

By studying art, science, the humanities, social science, and languages, the mind develops the mental dexterity that opens a person to new ideas, which is the currency for success in a constantly changing environment … Completing a broad liberal arts curriculum should enable a student to develop the conceptual, creative and critical thinking skills that are the essential elements of a well-exercised mind.

Lafley's claim coincides with the goals set forth in *The engineer of 2020: Visions of engineering in the new century*:

We aspire to engineers in 2020 who will remain well grounded in the basics of mathematics and science, and who will expand their vision of design through a solid grounding in the humanities, social sciences, and economics. Emphasis on the creative process will allow more effective leadership in the development and application of next-generation technologies to problems of the future….

We aspire to an engineering profession that will rapidly embrace the potentialities offered by creativity, invention, and cross-disciplinary fertilization to create and accommodate new fields of endeavor, including those that require openness to interdisciplinary efforts with *nonengineering disciplines* [emphasis added] such as science, social science, and business. (National Academy of Engineering [55, pp. 49–50])

At least some US students agree with the value of *nonengineering disciplines*. A survey by BestColleges found that 27% of STEM (Science, Technology, Engineering, and Math) students listed soft/professional skills as "the most valuable benefit" of their degree [40].

Among the academics I consulted in the UK, there is some debate about the role of general studies in engineering education. The Bologna Process imposes time limits on bachelor's, master's, and doctoral coursework. Some argue that general studies are not essential for engineers and that they take up valuable time required to achieve outcomes for specialized courses such as engineering. However, others maintain that general studies are an important part of an engineering education and should be a mandatory part of all engineering degrees. A report from the UK National Foundation for Education Research (NFER) states that "Problem solving/decision making, critical thinking/analysis, communication, collaboration, creativity, and innovation are transferable skills which will be in high demand in the next 15 years and beyond as technology becomes more embedded in the workforce" [73].

Similarly, the Royal Academy of Engineers argues that,

Engineering businesses now seek engineers with abilities and attributes in two broad areas—technical understanding and enabling skills. The first of these comprises: a sound knowledge of disciplinary fundamentals; a strong grasp of mathematics; creativity and innovation; together with the ability to apply theory in practice. The second is the set of abilities that enable engineers to work effectively in a business environment: communication skills; team-working skills; and business awareness of the implications of engineering decisions and investments.

It is this combination of understanding and skills that underpins the role that engineers now play in the business world, a role with three distinct, if inter-related, elements: that of the technical specialist imbued with expert knowledge; that of the integrator able to operate across boundaries in complex environments; and that of the change agent providing the creativity, innovation and leadership necessary to meet new challenges. (Royal Academy of Engineering [66], p. 4])

Most engineering educators agree that these types of knowledge and skills are necessary to help engineers be successful in their careers. However, methods for incorporating this information, typically learned in general education classes, into an already packed programme are still evolving. (This topic will be discussed further in the second essay of this series.)

Format
Although there are exceptions, most universities in the UK use a block module delivery system. Students begin a course and proceed from term to term as a cohort comprised of all students in a course who entered in the same term, taking all required modules together simultaneously. This format can facilitate group study, team projects and research, and community building, but it precludes most individualization (at least for the first two years) and makes part-time study extremely difficult. Part-time students are almost non-existent in the UK, a fact that has important implications for broadening participation. (This topic will also be discussed further in the second essay of this series.)

In the US, students choose their own class schedules and can choose any class in any semester as long as the course is offered and they have met the course prerequisites. This flexibility facilitates university study for students who must work or meet other obligations while attending school. In fact, over the last decade, an average of 38.3% of undergraduate enrollees in the US were part-time students [56].

Delivery
In the UK, the most common method for delivering information is the formal lecture (Hylton, & Hylton [39], p. 5). This format is based on educational traditions dating back to the UK's earliest institutions. A student generally has two to three hours of lecture in one or two sessions each week per module. Lectures often address a large group of students in a theatre setting with limited interaction between students and the lecturer. Lectures are supplemented by more interactive group tutorials or seminars in smaller groups, along with self-directed study. The university categories are helpful here. The lecture format is more common in the more traditional universities that make up the Russell Group. *Post 92* institutions are much more likely to use a variety of educational methods.

The Royal Academy of Engineering (2014) notes the limitations of the lecture as a teaching strategy in its 2014 publication *Thinking like an engineer: Implications for the education system*:

> The lecture still dominates as a teaching method in higher education and project work does not have "sufficient disjuncture to cause the learner to exercise reasoning and make judgments." ... Projects are still guided too much by the teacher or lecturer and do not encourage problem-finding and improving in particular, as students search for the "correct answer." The innovations are often limited to one course or module within a course, rather than having been adopted by the whole department and the departments tend to operate in a silo mentality. (Royal Academy of Engineering [66, p. 40])

In the same report, Professor Iain MacLeod argues that project learning should be increased: "To achieve a twenty-first century standard of engineering the use of much greater proportion of project learning than in traditional curricula is essential" (as cited in Royal Academy of Engineering [66, p. 40]). Thomas Litzinger agrees: "Why, when compelling evidence exists for the effectiveness of methods such as peer learning and inquiry-based learning in science education, have such methods not seen greater adoption in engineering?" (as cited in Royal Academy of Engineering [66, p. 54]).

The report exposes concerns about educational methods regarding engineering in the UK educational system as a whole, beginning in the early pre-university years:

> So, for example, if you want 19-year-olds who can think for themselves, solve problems with others and persist in the face of difficulty, then you will not give them pre-packed topics, individual tasks and problems which are well within their comfort zone. Instead you will invite them to take a role in designing their own learning, train them in the different roles and methods needed in successful group work and reward them for pushing themselves hard, making mistakes and bouncing back to do even better as a consequence.

> With regard to engineering education, our working hypothesis is that the current system, at a fundamental level, uses teaching and learning methods which tend only accidentally to develop engineers. (Royal Academy of Engineering, [66, p. 6])

In the US, content delivery methods are more varied, especially post-COVID. The lecture is still used, but it is far less common than in the UK [39, p. 5]. Teachers at every level are encouraged to use teaching strategies that appeal to students with various learning styles. The following methods are among the most common:

- Collaborative learning: Working in groups, students ask questions, debate, and discuss course topics. Benefits include learning skills such as communication, active listening, teamwork, and social interaction.
- Technology-based learning: Leveraging digital tools such as smart boards and webinars to enhance flexibility in learning.
- Flipped classroom: Students learn independently, then practice using the information independently or in groups, virtually or in-person. This method reverses the traditional practice of having students learn the material in class, then practice on their own.
- Socialization: Post-COVID, educators have realized the importance of socialization, even in online learning. Greater emphasis is being placed on interactive discussions

that include socialization to promote communication and keep students motivated to learn.
- Gamification: Gaming techniques have been found to be as effective for universities as they are for pre-university student learning. Teachers are increasingly taking advantage of online gamification platforms to creatively engage students.
- Combinations of teaching strategies that include techniques such as virtual learning environments that blend audio and video with online and face-to-face learning in collaborative ways [38].

Compared to UK requirements, students in the US must attend more classes and complete many more graded assignments that contribute to their course grades [39, p. 5]. Classes include discussion and interaction between students and the professor. My UK colleagues were surprised to learn that in the US, class participation is frequently part of a student's course grade.

Engineering education programs in the US also tend to place greater emphasis on teams and teamwork to help students develop skills for understanding themselves, other team members, and the contexts each team member brings to the team. Students participate in multiple team projects with different team members. Team projects are also used in the UK, but they are not the major learning components that they are in the US.

Assessment

Assessment is largely summative in the UK. Often, 70–100% of a student's grade in a module is based on the final exam. Students have little to no required (formal, marked) coursework; they work with an academic mentor as they develop and conduct their research. In the US, assessment is largely formative. Multiple components (e.g., assignments, papers, projects, and class participation) make up about 50–70% of a student's final grade, including sub-exams that make up about 30–40% of that amount. The final exam is typically weighted at about 20–30%. Grading scales may vary from one university to another. Comparing UK and US assessment can be difficult since assessment procedures and scales vary widely. Table 6 describes the most common assessment scales in the UK and the US [69].

Some US universities or departments have adopted standardized formulas for course grades. Still, in most cases, grading formulas are determined by the individual professor, and professors can adjust grades based on the difficulty of the task and overall student performance. In such cases, a grade of D could be given for a 40% score, but that decision is subject to the professor's discretion for that course. Adjustments are made in the UK as well, generally based on the grade distribution for an assessment. However, those adjustments must be identified and justified to external examiners in a much more rigorous and formal regulation process.

After teaching two terms at a university in the UK as part of a Fulbright Visiting Scholar Exchange program, Purdue University professor Pete Hylton had the initial

Table 6 Comparison of most common module/course grading scales in the UK and the US

	UK	US
90–100%	A—First-class honors (1st) Excellent to outstanding	A—Superior
80–89%		B—Excellent
70–79%		C—Average
60–69%	B—Upper second-class (2.1) Good to very good	D—Below average
50–59%	C—Lower second-class (2.2) Satisfying	F—Fail
40–49%	D—Third class (3rd) Sufficient	F—Fail
0–39%	Fail	F—Fail

impression that "the assessment of engineering students in the UK was much less rigorous in terms of the expectations for mastery and the level of effort required for a course." He writes,

> When UK college students heard the level of expectation for exams, projects, and required homework assignments at USA universities, they were totally aghast at the amount of work required. By contrast, American students would probably have heart failure if they learned that some UK schools give only one exam, and that it may come at the end of the school year, which might be as long as 5 months after the end of the classroom sessions for the material being tested. Hylton and Otoupal-Hylton [39, p. 9]

An initial comparison of percentages and letter grades may seem to suggest that American universities have higher grading standards. However, since percentages and grades reflect the expectations and performance of different education systems (including module vs. course), any comparison of grading scales leading to conclusions about grading rigor would be speculative without a more systematic approach to analysis.

Emphasis on Sustainability

The societal and political issues of climate and sustainability have spilled over into education. My interviews with academics in the UK revealed that they are puzzled by the seeming reluctance of the US to embrace climate sustainability and by the level of skepticism on climate change—or at least on the causes of climate change—by conservative political forces in the US. A 2023 report from NASA concludes that 97% of scientists "agree that humans are causing global warming and climate change" [51]. Individuals in the UK and the US are less certain than the scientists, but Americans have the least consensus. The Pew Research Center reports that "54% of Americans view climate change as a major threat," a *decrease* of 5 percentage points since 2019. Climate change is a politically divisive issue: 78% of Democrats view "climate change as a major threat" (down 10 points from 2020) compared to only 23% of Republicans (down 7 points from 2020).

Perhaps, then, it is not surprising that Americans rank climate change as number 17 in a list of the top 21 national issues [79].

In the UK, public concern is significantly higher. A 2022 survey shows that 82% of UK residents were either "very concerned" (38%) or "fairly concerned" (44%) about climate change [67]. *The engineer of 2020: Visions of engineering in the new century* states that engineers need "practical ingenuity" to find solutions and provide leadership on issues like climate change:

> Yesterday, today, and forever, engineering will be synonymous with ingenuity—skill in planning, combining, and adapting. Using science and practical ingenuity, engineers identify problems and find solutions. This will continue to be a mainstay of engineering. But as technology continues to increase in complexity and the world becomes ever more dependent on technology, the magnitude, score, and impact of the challenges society will face in the future are likely to change. For example, issues related to climate change, the environment, and the intersections between technology and social/public policies are becoming increasingly important. By 2020, the need for practical solutions will be at or near critical stage, and engineers, and their ingenuity, will become ever more important. (National Academy of Engineering [55, p. 55])

Similarly, the Royal Academy of Engineering (RAEng) believes that engineers have a responsibility to play a leading role in the transition to a net zero carbon economy and to use their skills and knowledge to help address climate change. They are calling on engineering to "step up and play its role" in building a sustainable future [49].

A policy paper from the UK Department for Education (2022) mandates a sustainability strategy that applies to schools at all levels (including higher education). The strategy "requires the education sector to play its role in positively responding to climate change and inspiring action on the international stage" [22]. My discussions with academics in the UK revealed that climate change sustainability is thoroughly integrated as a design criterion in engineering projects/designs.

Sustainability policies in the US lag behind. The Climate School at Columbia University states that "In the US, more than 86 percent of teachers and 84 percent of parents support climate change education in schools." However, in early 2023, "two measures that would have supported climate change education died in Congress. There is still no national consensus about the importance of climate education, and the U.S. does not have national science standards" [12]. In 2020, the National Center for Science Education and the Texas Freedom Network Education Fund graded all 50 states plus the District of Columbia on their implementation of the Next Generation Science Standards (NGSS) developed in 2013:

> A bare majority — just 27—of the 50 states and District of Columbia have standards that earned a B+ or better for how they address climate change. Those 27 include the 20 states and DC that have adopted the NGSS. Of the remaining 24 states, 20 earned no better than a C+. Ten of those received a D or worse, and they include some of the most populous states in the

country, such as Texas (F), Florida (D), Pennsylvania (F), and Ohio (D). Six states received a failing grade overall. (*Making the grade* [46])

As engineering educators, we must rate our own performance in sustainability education and, as the report from the Royal Academy of Engineers states, "step up and play [our] role" in building a sustainable future [49].

Students

Entry Requirements
After completing secondary school at age 15–16, students in the UK who plan to attend a university spend two years in a preparatory study of 3–4 subjects that are critical to success in a university course. This study culminates in A-level exams, a single exam somewhat comparable to an ACT or SAT exam in the US except that instead of testing a broad range of subjects, it focuses on subjects relevant to a specific course. Students are allowed to repeat the A-level exam, and academics use A-level scores to evaluate which students they will accept into their course. Requirements for admission vary somewhat among UK universities and courses. Typically, at more selective universities, entry into a mechanical engineering course requires two As and a B, and one of the As must be in calculus. Some Russell Group universities require 2 A+ and one A.

The pre-admission testing process is quite different in the US. Until recently, students were required to submit scores from one or both of two college entrance exams: the American College Test (ACT) or the Scholastic Assessment Test (SAT) as part of their application. The ACT tests four subject areas: English, math, reading, and science, plus an optional essay to test writing proficiency. The maximum score on the ACT is 36; the average score is 21 [87]. The SAT test assesses math, evidence-based reading, and writing (with subsections in reading and writing/language). The maximum SAT score is 1600; the average score is 1060 [87]. Percentile scores are based on the average of all students who take the test. Many institutions will accept ACT scores near the national average of 21 and SAT scores in the 950–1050 range. Highly selective schools usually require scores between 32 and 36 on the ACT or in the 75th percentile and above on the SAT [48, 50].

Testing is a multimillion-dollar industry for the College Board (SAT) and the Iowa Testing Company (ACT) who own and administer the tests and for businesses that provide preparation guides, classes, and private tutors. To take the SAT, each student pays a registration fee of 55 USD (plus possible change and/or late registration fees of 25 USD and 30 USD) [88]. Registration for the ACT costs each student 63 USD; the cost increases to 88 USD if the student includes the optional writing portion of the test [87]. Before the test, many students pay for test preparation courses or private tutoring. Students who are unhappy with their first results often re-take a test, and many take both tests to see which gives them the better score.

Over the last decade, however, universities in the US have begun to question the usefulness of these exams as criteria for admission. Critics cite research showing that the tests are less effective at predicting higher education success than other measures, such as recommendations from teachers and/or guidance counselors, a student essay, and a student's high school GPA. (UK schools have no GPAs.) The tests also perpetuate inequalities between students from wealthy and low-income families. Schools in affluent communities frequently offer test preparation courses at no cost to students, while schools in less prosperous communities cannot afford to do so. Less affluent students usually do not have the option of taking both tests or repeating the test(s). The test developers have also been accused of selecting questions that perpetuate racial inequalities.

The ACT and SAT tests—once mandatory for college admission—are rapidly becoming obsolete. More than 80% of US colleges and universities no longer require a college entrance exam. Test-optional institutions have doubled since 2020, and 78% of those institutions have extended "test-blind" and/or "score-free admission" policies through at least fall 2024 [14, 57].

However, some top schools are reinstating the test requirement. In 2022, MIT announced that it would again require standardized test scores. Dartmouth and Yale joined them in early 2024. Yale claims that its internal study reveals that "rejected low-income students who omitted SAT scores but scored in the 1400 s would otherwise have been admitted. The school would have taken those scores into account had the students reported them" [60]. Testing remains controversial, but for fall 2025, 80% of 4-year colleges and universities will not require test scores [14].

Academics I interviewed in the UK were quite surprised to learn that US students take a standardized test that is not specific to the students' intended course of study. They will likely be even more surprised to learn that pre-admission testing is rapidly disappearing from the admissions process. These entrance testing differences are consistent with design differences between focused and broad educational programs.

Universities in both the UK and the US want to attract students who will be successful. They differ, however, in their methods for identifying those students and in the information available for use in making decisions. UK universities focus on summative assessments and look for students who are prepared to be successful in their courses. US universities look at formative assessments. They look for students who have a broader range of knowledge, essentially showing that they have learned how to learn, but they also consider a student's high school grades in subjects relevant to the student's degree choices. The UK method may weed out many students who are underprepared to succeed in their chosen course. This is an important criterion since the siloed nature of UK modules makes changing a major very time consuming and costly. The US method is likely an extension of the CTE opportunity to explore—and reject as well as choose—options for academic and career directions [13].

Both UK and US universities have a pathway available for inadequately prepared students. In the UK, students who do not meet the admissions criteria of A-levels can take a foundation year course. The foundation year does not count toward a university course, but if students pass, they are often guaranteed admission into the university that teaches the foundation year. Sometimes universities will also accept another university's foundation year for admission. Students may also sit out a year or attend a local college before retaking A-level exams. Internship is another option; after a year-long internship, students may retake A-levels and reapply for admission.

In the US, inadequately prepared students can attend a community college for two years to improve their GPA before applying to university, or they can enter university as an undeclared major. Unlike UK colleges and the foundation year, US community colleges have course credit pathways to universities that reduce the time to graduation, sometimes by as much as two years.

These secondary pathways to enter a course or degree program reinforce the difference between the emphasis on focused (UK) and broad (US) programs. They also have significant implications for the goal of broadening access and attracting students from underserved populations.

Admission

In the UK, students apply to university through a centralized system, the Universities and Colleges Admission Service (UCAS). Students submit only one application and select up to five courses, usually at different universities. Academics review applications and decide which students will be admitted. A student's having been accepted into one course does not transfer to a different course (e.g., mechanical vs. biomedical engineering). Because all modules are course-focused, a student who wishes to change courses would have to reapply to a different course and start over. Module credits rarely transfer from one institution to another, so course and university changes can be costly in both time and money.

In the US, a few universities require students to apply to a specific program, but most students apply to a university rather than to a specific department or degree program. Students apply separately to as many colleges and universities as they wish and choose from the offers received. Most institutions have an Admissions Office that handles all student applications and makes university acceptance decisions. Faculty in university departments generally have little to no input on student acceptance.

Change of Major

US students beginning university studies may or may not know which degree they wish to pursue. One survey found that 51% of US students applying to college were uncertain about their desired career [58], so it is not surprising that 20–50% of students begin college as an "undeclared" or "undecided" major [72]. Since the first 1–1.5 years of a baccalaureate degree consist primarily of general education classes, in most cases, students can take time to decide on a major. Approximately one-third of undergraduates in four-year colleges change their major at least once, and about 10% change more than once [40]. Yet, it is important to note that changing majors may require additional time and cost in some majors, especially engineering programs.

Interestingly, a *U.S. News and World Report* article actually recommends an undeclared major for some students, specifically in engineering programs:

> If you have a competitive college concentration in mind, your academic profile—your GPA, in other words—will be key. If your high school GPA does not reflect your full potential and you would like to use your first year of college to correct this issue, it likely makes sense to apply as an undeclared major. This is a particularly good idea if your high school GPA is weak in the major's core field or fields.
>
> Engineering is one common major where this strategy may apply. Because engineering offers strong career prospects, it can be a popular concentration. Thus, universities can be highly selective in which applicants they accept to their engineering schools. If you lack a history of high school success in science and math classes, it may be best to take college-level courses in STEM fields before you apply to this major. (Holmes and Duda [37])[5]

While this advice may be applicable to most universities in the US, it runs counter to the UK model. Since no A-level exams specifically test engineering, students take exams in math and science, subjects that many educators believe are essential to success in engineering. Since the UK requires application to a specific course, students cannot apply to or enter university as an undeclared major.

The US admissions process allows flexibility. The UK process provides pre-admission direction and preparation. Both flexibility and direction can have advantages—or disadvantages—for a student. This is an important area for discussion to help us learn how best to prepare, motivate, and train students as well as how to broaden access to engineering education.

Students' Union

Nearly every UK university has a students' union. Students' unions are independent of their universities, and most are members of the National Union of Students (NUS), representing all UK students [47]. The first students' union was founded in 1873 at King's College London to create unity among students on campus [89]. Unity and socialization of students remain core functions of today's students' unions.

But the role and impact of students' unions have evolved to include a broad range of student issues within and beyond the university context. NUS now has a seat on the boards of the Higher Education Funding Council for England (HEFCE), the Office of the Independent Adjudicator (OIA), and the Universities and Colleges Admissions Service (UCAS) [52]. Most students' unions also have representation on the local University Council (similar to the Regents or the Board of Governors in US universities). These roles give NUS, representing all students' unions, a voice in higher education admissions,

---

[5] Acceptance into a specific major varies among US universities. Some universities require a minimum GPA in relevant courses before acceptance into an engineering program.

funding, and student complaints. NUS and its member students' unions have actively lobbied for and/or protested against caps on student loans, interest rates, the inadequacy of maintenance loans to cover the cost of living, insufficient funding to support student mental health, and other issues affecting students [53]. For the last five consecutive years, NUS has supported university staff in strikes protesting "government cuts to education and slashes to workers' rights" [54].

Students' unions are also active in protecting the consumer rights of students. The Office for Students (OFS) states that "Students at universities and colleges in England are covered by the principles of consumer protection, and a range of law and guidance applies" [62]. If students believe that their consumer rights have been violated by a university, they can engage the students' union to advocate on their behalf. Consumer protection laws require higher education providers to protect the rights of students in the following ways:

- "Ensuring that students are given up front, clear, timely, accurate and comprehensive information" about courses offered, course structures, fees and costs, terms and conditions, rules and regulations, and any anticipated changes to this information.
- "Ensuring that terms and conditions between HE providers and students are fair." Terms must be explained in clear, unambiguous language, and any terms that are "important or surprising" must be highlighted to call attention to their significance.
- "Ensuring that HE providers' complaint handling processes and practices are accessible, clear, and fair to students". (*UK higher education providers—advice on consumer protection law* [81, pp. 5–9])

A students' union has formidable power to call an institution to account, ensuring that a student complaint receives a hearing and a fair resolution.

Institutions in the US have less accountability than those in the UK. Faculty in the US are much more independent and have greater freedom to develop and teach courses as they see fit. Students' unions do exist in US universities, especially among graduate students, but they are far less common than in the UK. However, unions are becoming more prevalent on US campuses. According to *Inside Higher Ed*, "In the past three years, the number of recognized undergraduate unions went from one—the University of Massachusetts at Amherst's 20-year-old union of Resident Assistants and Peer Mentors—to over a dozen" with about a dozen additional campaigns to unionize underway [42].

Students' unions in the US have primarily focused their attention on issues regarding work conditions, but they have also accomplished some institutional changes. For example, a clash between the graduate student union and the administration at Temple University led to the university president's resignation [42]. While many institutions have a peer evaluation process for student grievances, the administration retains the power to make decisions.

Faculty

Faculty Development (Teaching)
Academics I consulted in the UK believe that effective teaching is sometimes highly valued, especially at less research-intensive universities. (This observation may be highly influenced by interviews with academics who were focused on teaching within engineering education.) Consumer protection laws and students' unions contribute to this valuation: as consumers, students can file a complaint if module content or instructional methods vary from published specifications or if they feel that the quality of teaching is substandard. Many universities include training students to be teachers as part of their PhD courses, often in schemes aligned with the Professional Standards Framework (PSF) for Teaching and Supporting Learning in Higher Education. The PSF originated in England and has been used by universities around the globe. Membership and use of the PSF is voluntary.

The PSF outlines three "dimensions" intrinsic to effective teaching: Professional Values, Core Knowledge, and Areas of Activity. The PSF (updated in 2023) emphasizes "the effectiveness and impact of teaching, the context in which the teaching takes place, and … more inclusive approaches [that] ensure all learners feel respected, valued and have equity in opportunity to succeed." It also places "greater emphasis on digital/technology, professional values, support for students, and collaboration" [61]. Administration of the PSF program falls under the Higher Education Academy/Advance HE, which can award accreditation to "continuous professional development (CPD) programmes for teaching and learning which are delivered by higher education providers around the world." Academics who teach at PSF member universities can be awarded four levels of fellowships to recognize the quality and professionalism of their teaching [74]. Some member universities require academics/faculty to achieve PSF goals for promotion.

Despite agreement on the importance of teaching, UK academics tend to believe that teaching is often undervalued compared to research (especially in terms of promotion) and that "there is a large discrepancy between the value they perceive to be given to teaching and the value they believe it should be given. The largest discrepancies between the perceived and the desired emphasis given to teaching occur in more research-intensive institutions" [63].

Most faculty in US universities agree that quality teaching is important and should be highly valued. However, they share the concerns of their British counterparts that the quality of teaching is less valued than it should be, especially in comparison to research. The Harvard Business Review reports that

> the quality of universities, at least as judged by research excellence tables, is predominantly based on research rather than teaching. In many top institutions, teaching can be seen as a distraction from publishing and getting research grants. Top faculty are attracted not just with higher salaries, but also with more freedom and a lower teaching load. In return, they will

publish research prolifically and bring in grant income while leveraging graduate students to do their teaching instead. (Chamorro-Premuzic and Franiewicz [10])

Of course, these generalizations have exceptions in both the UK and the US. The question remains, however, whether quality teaching is adequately rewarded. Most teaching-track faculty believe that teaching and research should be valued equally and should be rewarded with greater parity through promotions and salary increases and that universities should provide ongoing faculty development resources to improve teaching.

In the past, academic tenure has been one of the important ways universities rewarded quality in both research and teaching. Prior to 1988, tenured faculty enjoyed greater job security, especially in the UK, along with higher salaries and greater academic freedom (in the US). The Education Reform Act of 1988 ended tenure in the UK. In its place, academics are hired on either permanent or fixed-term contracts. Tenure survives in the US, but it is under attack both from university hiring policies and from state governments. Data from the National Center for Education Statistics (NCES) (displayed in Fig. 3) reveals that between 1987 and 2021, the percentages of contingent and part-time faculty have increased while the percentage of tenured faculty has decreased [15].

Quality Control/Assurance Versus Academic Freedom
While academic tenure has traditionally been part of a reward system, its primary purpose—at least in the US—is the protection of academic freedom [75]. In this realm, education in the UK and the US are very different.

In the UK, "External examiners advise and, where necessary, challenge the institution on how their students are achieving threshold academic standards" [26]. Accountability

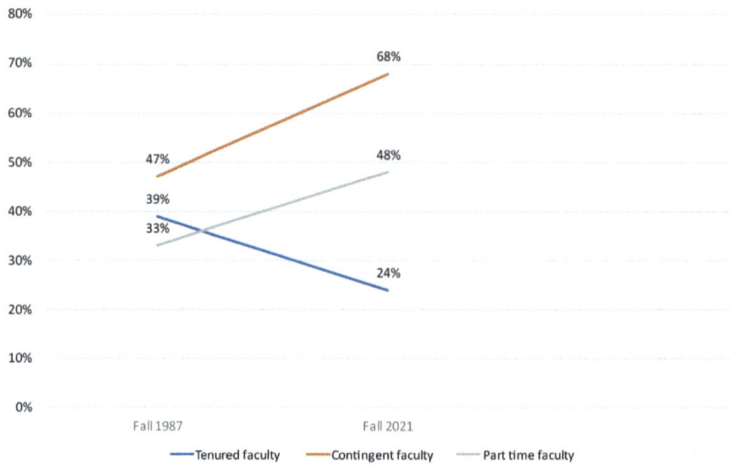

Fig. 3 Changes in academic employment, Fall 1987–Fall 2021

to external examiners, along with government oversight, accreditation policies, and the consumer protection powers of students' unions create a highly structured environment that constrains how academics develop course and module content, deliver content, and assess student work. When academics in the UK teach a module, they begin from a set of extremely detailed course policies and instructions that prescribe the module content, teaching methods, and assessment procedures. This information is also available to students. Changes or deviations are violations of consumer protection laws. Thus, students keep faculty accountable to teach the module as prescribed.

At some of the universities I visited, I had opportunities to observe classroom instruction and talk with students. I asked students in one classroom what would happen if their academic came to class one day and said that their lab would now count as 15% of their grade, and their exam would count only 15% instead of thirty percent. At least 1/3 of the students raised their hands and said the academic could not do anything that would violate the module's specifications. Changes to a module would have to be proposed to a committee for review and approval, a process that would likely take at least a year. My UK colleagues said that some areas have a small amount of "wiggle room," but that content and assessment are tightly controlled to maintain and assure quality in each module of a course.

These colleagues were quite surprised when I described the experience of faculty in the US. Typically, faculty develop and/or teach a course based on the brief description in the university catalog. Below is an example of a catalog description for CE 3336, a required course for Civil Engineering majors at UTEP. *BS in civil engineering* [4]

CE 3336. Civil Engineering Materials.

Civil Engineering Materials (2–3) Properties of civil engineering materials, measurements and test methods, relationship of properties to performance; their structure and behavior: relationship between structure and behavior. Prerequisite: Junior standing in Civil Engineering.

Department: Civil Engineering

3 Credit Hours

5 Total Contact Hours

3 Lab Hours

2 Lecture Hours

0 Other Hours

Prerequisite(s): (CE 2334 w/C or better)

This brief description is the initial foundation of course development or teaching. Of course, a new faculty member would have access to syllabi from previous instances of the course and would likely consult with other departmental faculty; but then, within the workings of the department, the faculty member would develop or teach the course and assess student work as they saw fit.

Accreditation procedures along with historical and cultural contexts contribute to the differences between the UK and the US in module/class development, teaching, and assessment. UK accreditation is based on a set of prescriptive standards and procedures that are outlined in the UK Quality Code for Higher Education and are used to assess a university's processes to assure quality [76]. Quality assessment is overseen by the UK Standing Committee for Quality Assessment (UKSCQA), the Quality Assurance Agency (QAA) and the Office for Students (OFS). The Quality Code describes itself as a "reference point for effective quality assurance" [76, p. 1].

US engineering accreditation is far less prescriptive. Engineering programs are accredited by the Accreditation Board for Engineering and Technology (ABET), a non-profit entity comprised of a federation of 35 professional and technical *member societies* representing the fields of applied science, computing, engineering and engineering technology. ABET accreditation is a voluntary process that ensures that engineering programs meet the quality standards of the profession. While ABET does have eight general criteria that each accredited engineering program must demonstrate (Students, Program Educational Objectives, Student Outcomes, Continuous Improvement, Curriculum, Faculty, Facilities, and Institutional Support) along with various program-specific criteria, it begins from a non-prescriptive, program-based perspective. Instead of basing assessment on prescriptive quality standards, it originates from the questions, "What are your program's goals? What are you trying to achieve?" It then assesses an institution's program based on the how well the program achieves those goals with a strong emphasis on continuous improvement. ABET accreditation is valid for six years. After that time, the program must undergo another peer review process to maintain its accreditation.

This difference in accreditation procedures raises the critical issue of how much freedom individual academics should have and what safeguards should be imposed to assure quality outcomes for students. The UK model is based on the belief that precise standards and specifications serve as safeguards to ensure quality. By contrast, in the US, such prescriptive standards and specifications would be perceived as detriments to creativity and innovation and as threats to professors' freedom of expression. The absence of such prescriptiveness equals academic freedom. Many US faculty would readily admit that academic freedom is risky; some bad things will happen. But most professors in the US believe that the risk is worth taking since it allows greater creativity and individualization for both faculty and students and protects faculty from being fired for individualistic expressions or actions.

These attitudes toward prescriptiveness and academic freedom grow out of the different historical and cultural contexts in which UK and US universities operate. The UK has a long and storied history, beginning in the ninth century. Centuries of monarchy—along with proximity to and alignment with other European nations—have shaped the UK's cultural identity and traditions. The US did not begin to exist as a nation until the eighteenth century—nearly a thousand years later. The US was born out of a belief in the supremacy of the individual and his or her right to freedom and independence. US settlers passed those values on to their descendants, and they persist in the culture and attitudes of US citizens.

The British have a historical and cultural bent to show respect for society and to base their values on the community. People in the US tend to feel freer to show disrespect, to resist what they perceive as restrictions from authorities, and to value individualism. It is not surprising that these differences show up in our educational systems because they are foundational to our identities as people groups and nations. Most US faculty and institutions would chafe under the prescriptive structure in the UK. After all, the US founders basically thumbed their nose at the king, declared themselves free and independent, and fought a war to prove it. Most UK academics would reject the risks of academic freedom. Educators in the UK and the US share the goal of providing quality education for students, but our definitions of that quality and our means of assuring it can be quite different.

External Factors

Industry-University Partnerships
Industry involvement in education is much greater in the UK than it is in the US. Many UK university courses include internship opportunities for students. These range from short-term opportunities, lasting only a few weeks or months, to a "sandwich year," which lasts for nine months to a year and generally includes academic requirements. The sandwich year gives students a professional development opportunity along with job experience that increases their employability. Along with holidays and other benefits, students are paid for their work during the sandwich year. Salaries average £18,670 (about $23,300 in US dollars), but students must also pay university tuition fees. Fees vary from one university to another but are generally about 20% of the £9250 tuition rate [17]. My UK colleagues stated that in some institutions, about 85% of students do a sandwich year; in others, only about 10%.

A UK Partnerships Consultant writing for the *Higher Education Policy Institute* (*HEPI*), states that the number of students enrolled in sandwich year programs (not internships) has "increased by 59% between 2009–10 and 2017–18." However, that increase is not sustained in the number of students who complete a sandwich year. Despite more universities offering sandwich years, only 8.2% of students in the 2005–2006 cohort and only 8.6% of the 2014–15 cohort completed a sandwich year placement [33].

This data is surprising when compared with employment and salary statistics. Table 7 presents data from the Department for Education [21, p. 31] showing that in each measured time period, graduates whose course included a sandwich year have a higher percentage of further study, sustained employment, or both.

Table 7 Graduates in further study, sustained employment, or both

Mode of study	Years after graduation							
	One		Three		Five		Ten	
	Number	%	Number	%	Number	%	Number	%
Full-time	231,375	87.3	238,940	86.2	215,185	85.2	176,530	82.9
Sandwich	18,545	89.0	15,150	87.1	13,035	85.3	14,865	83.5
Part-time	31,055	85.9	32,335	85.6	30,565	84.3	28,965	80.3

Table 8 Median earnings by mode of study

Mode of study	Years after graduation							
	One		Three		Five		Ten	
	Number	£	Number	£	Number	£	Number	£
Full-time	145,905	19,200	167,085	22,900	154,755	25,600	128,295	30,200
Sandwich	13,715	24,000	11,335	28,900	9890	31,800	11,385	36,600
Part-time	19,955	23,100	22,045	24,900	21,220	26,200	19,605	29,300

Table 8 displays additional data from the same report [21], showing that a sandwich year also leads to higher income.

In each time period, graduates whose course included a sandwich year earned higher median incomes. Grey [33] notes, however, that this sector-wide data may be somewhat misleading because "Many of the courses with higher placement take-up rates, such as engineering disciplines, have stronger labour markets and lead to higher salaries on average across all graduates."

Not every sandwich course leads directly to a job with the sandwich year employer. I spoke with a few students who had spent their sandwich year working at an automotive company. One told me, "I've seen enough. I no longer desire to work for the automotive industry." In fact, my UK colleagues told me that overall, only between 1/3 and 1/2 of students return after graduation to work at the company where they spent their sandwich year.

The US has no specific sandwich year program, but students have internship opportunities to gain work experience. These are much less formalized than the UK's sandwich year, and the responsibility for finding and obtaining an internship lies much more heavily on the student. Internships tend to be much shorter than the sandwich year, lasting a summer, a semester, or even just a few weeks. Fennell [27] reports that "21.5% of college students do an internship," and the "average internship lasts for 18.3 weeks."

About 60% of students are paid for their work during an internship. Paid interns receive 69% more job offers than unpaid interns and typically earn higher starting salaries, but about 40% of internships in the US are unpaid [35, 41]. The US Department of Labor cites the "primary beneficiary test" as a determining factor on whether a company is required to pay interns. The test evaluates factors such as expectation of payment, similarity of internship training to an educational environment, job adjustments based on the intern's education and school calendar, amount of time working that is perceived as benefitting the learner, relationship to jobs of existing workers, and an understanding that the internship does not entitle the intern to a paid position [80]. Non-payment impacts many students' ability to take an internship. According to the National Survey of College Internships, about 40% of students who desired an internship were unable to apply because they had to earn an income [41].

UK Student Funding and Need for International Students
The British government provides funding for each "home" (British) student to attend university for up to four years (which may include a preparatory year or an integrated master's degree). Funding usually has two parts: a loan to cover tuition fees and a maintenance loan for living costs. The government caps the loan amount, and that cap sets the tuition fee. If an institution were to charge a higher tuition fee, students would not be eligible for the loan. Because of the equity culture in the UK and the cap on loans/tuition fees, all universities charge the same amount for tuition fees. Currently, the maximum loan for tuition fees in England (as well as the maximum allowed tuition cost) for home students is 9250 GPB per year (approximately 12,000 USD at current conversion rates). Maintenance loans are also available to support cost of living. Maintenance loan amounts vary based on family income.

Government loans become part of a student's financial record. Interest accrues, but no repayment is required until the student graduates, is employed, and reaches a certain level of income. As of September 2023, that income threshold was reduced from 27,295 GPB to 25,000 GPB [86]. Once a worker reaches the income threshold, the employer takes the loan payments directly out of the former student/now employee's salary. Currently, payments must be made for 30 years, after which the loan is written off. However, that repayment period is expected to increase to 40 years in the near future [9].

Loans for tuition fees are paid directly to the university when a student matriculates. However, inflation has affected the "real" value of tuition fees received. Weale [86] writes that "according to Universities UK (UUK), which represents 140 higher education providers, the £9000 tuition fee for UK students, which was introduced for English universities in 2012 and was topped up to £9250 five years later, is now worth just over £6500 to universities.... the percentage of English universities reporting an in-year deficit increased from 5% in 2015–16 to 32% in 2019–20." Along with the £9250 per student tuition fee, the government provides some differential financial assistance to universities for each engineering student: "Universities receive an additional £1500 per student per year to deliver engineering courses, but ... it costs approximately £15,000 a year to deliver—leaving a deficit of £4500 to make up from elsewhere" [6].

The primary strategy for UK universities to offset funding deficits has been to attract and recruit international students whose tuition is not capped by the UK government. In the International Education Strategy 2020/2021, the British government set a goal to educate 600,000 international students in UK institutions by the year 2030. They exceeded that goal the same year with 605,130 international students, making up 22.0% of all UK students (15.7% of undergraduates and 39.1% of postgraduates) [84].

According to the British Council, tuition fees for international students average 22,000 GPB per year for undergraduates but can vary from 11,400 to 38,000 GPB depending on the university and the course of study. Postgraduate tuition fees average about 17,109 GPB per year but can range from 9000 to 30,000 GPB or much higher for the most expensive courses (e.g., medicine) [18]. On average, about 1/5 of the funds received by UK universities come from international students; at some universities, international student fees add up to more than a third of a university's funding [30].

The tuition fee cap for home students coupled with new policies for international students have produced some unintended consequences. UK students have expressed concern about losing university places to international students who bring in more funds. Mark Corver, Managing Director and Co-Founder of DataHE states, "I know [the universities] would like to take more UK students, but the finance directors will be saying it doesn't add up. In practice, they need a higher and higher proportion of international students each year" (as cited in Garcia et al. [30]). International students are also concerned, not only by rising tuition fees, but also by new limitations on international students. Beginning 1 January, 2024, dependents are allowed to accompany an international student only if the student is pursuing a Ph.D., doctorate, or research-based course or is sponsored by the government [70]. Weale [86] reports,

> Vice-chancellors are warning the current funding model for UK higher education is "broken" and have urged the government to review the system of tuition fees .... They have made clear that limits to overseas students announced last week on top of rising costs caused by inflation posed a serious risk to universities which would require more funding from government.

The Higher Education Statistics Agency reports that in 2021–22, 45% of postgraduate students and 15% of undergraduate students enrolled in UK universities were "non-UK domiciled" [36]. The international student population is much smaller in the US, comprising only about 4.6% of higher education students. Some scholarships are available for international students, but they can be difficult to obtain. Since international undergraduate students often pay higher tuition [59], they help to generate income for US universities, but on a much smaller scale than in the UK.

Licensure

Titles can be controversial and confusing in both the UK and the US. In the UK, the Engineering Council acknowledges that "Commonplace use of the word engineer in our language has evolved over many centuries. Hence anyone in the UK may describe themselves as an engineer" [71]. For example, during my time in the UK, I had a problem with my refrigerator, and the service company said they would send over "the engineer." In the US, the response would have likely been "the technician or repairman." The Engineering Council cautions, however, that professional engineering titles, such as Chartered Engineer, are protected by the Engineering Council's Royal Charter and Byelaws and that unauthorized use of professional titles may be prosecuted. To be a Chartered Engineer in the UK is a notable distinction. To earn that title, engineers in the UK must follow rigid standards described in detail by the Engineering Council. The standards fall under five categories:

A—Knowledge and understanding

B—Design, development and solving engineering problems

C—Responsibility, management and leadership

D—Communication and interpersonal skills

E—Professional commitment [78, pp. 8, 15, 31–39]

People in the US also take some latitude with the title engineer. There is no shortage of forum posts in which engineering students or people doing some sort of engineering work seek to call themselves an engineer (For example, see Alfaa123 [2]). Such queries generally receive negative responses. A position statement from the National Society of Professional Engineers (NSPE) specifies that "the indiscriminate utilization of the title 'engineer' by individuals who are not clearly professionally qualified as set forth in this position statement serves only to confuse the public." A "qualified individual" is defined by the NSPE as

1. An individual who is licensed under a jurisdiction engineering licensure law; or
2. An individual who has graduated from an ABET/EAC program; or
3. An individual whose degree has been evaluated as equivalent to an ABET/EAC program or one which has been approved by the National Council of Examiners for Engineers and Surveyors (NCEES) or a state or territorial engineering licensing board. (Committee on Policy and Advocacy [16])

The title Licensed Professional Engineer (PE) denotes the highest licensing level in the US. To earn that title, an engineer must meet the requirements listed above, pass the Fundamentals of Engineering (FE) exam, have four years of work experience as an engineer, and pass the Professional Engineering (PE) exam. Each state issues licenses with specific renewal requirements, and a PE must complete continuing education requirements for license renewal [24].

University Reputation and Tradition

Universities in both the UK and the US extol their reputations and traditions, but the long history of higher education in the UK makes these much more deeply ingrained. It is not surprising, therefore, that in the UK, reputation and tradition are powerful forces not only to attract students, but also to shape educational practices and impact policies. Given the comparatively brief history of the United States, I would surmise that very few US Americans understand or appreciate the impact of these centuries-long ideals.

The influence of Oxford and Cambridge, collectively known as Oxbridge, is inestimable. The number of MPs who attended Oxbridge slipped from 22 to 19% in the 2019 election, but that still leaves nearly 1/5 of MPs with ties to the longstanding ideals and traditions of these universities. Not surprisingly, 70% of Oxbridge graduate MPs represent the Conservative Party [45]. More broadly speaking, according to the Sutton Trust [77], 54% of MPs went to a Russell Group university [77].

The Ivy League, the closest US equivalent to the Russell Group of universities, is not as strongly represented in US government. Although eight of the nine current Supreme Court Justices attended Ivy League schools, only 5% of the current 118th US Congress members are Ivy League graduates [67]. This means that the UK has 10 times more (54% vs. 5%) representation of "elite" university representation in Parliament than the US has in Congress.

Oxbridge—and the Russell Group universities—exert a powerful influence not only in government but also in UK education as a whole. Institutions that want to change or innovate are challenged on two fronts. First, they encounter government regulations and oversight that prescribe tuition fees as well as content, delivery, assessment and even the amount of time that students can be required to spend. Second, they come up against centuries of tradition. Questioning that tradition is neither a popular stance nor an easy one since the UK is justifiably proud of its longstanding institutional traditions and reputation. Even the things—and sometimes *especially* the things—we cherish most can be barriers to innovation.

Engineering educators in the US have greater freedom to explore new educational methods and even definitions of engineering: to design by looking beyond "Does it work?" and ask questions like "Is it economically sustainable?" "Is it user-friendly?" "Does it have aesthetic appeal?" "Will it lead to customer satisfaction?"

Innovation can be difficult. It can be risky. We need to explore ways to question and evaluate our practices and outcomes to understand what innovations are necessary to equip students for the challenges of our rapidly-changing world.

Geography and Population

Finally, it is important to consider how land mass and population impact university cultures and influences. The UK is about 2.5% of the geographical size of the US and has a population of about 68 million, compared to over 340 million in the US [19]. The UK has an average population density of approximately 280 people per square kilometer (725 per square mile) compared to the US average of 37 per square kilometer (96 per square mile) [83]. Nearly 27% of England's population live in and around London with population densities ranging between 4213 and 10,936 per square kilometer (2618–6795 per square mile) [44].

The distance from the northernmost tip of Scotland to Oxford or Cambridge is within 805 km (500 miles). In the US, the distance from a city on the West Coast (e.g., San Francisco) to any of the Ivy League schools ranges from 4490 to 4982 km (2790–3096 miles), more than six times greater.

The fact that so many US universities are located much further from Ivy League schools, combined with the value placed on uniqueness and independence, allow the Ivy League schools to exert only minimal—if any—influence over universities across the US. Oxbridge, with its centuries-long history and traditions, combined with much nearer proximity to other UK universities, gives Oxbridge powerful influence over university operations and culture. Population density in the Oxbridge area extends that influence to UK politics and government. US universities simply have no level of influence comparable to Oxbridge.

Conclusion

The significant differences between the US and UK systems of engineering higher education play a major role in the philosophies and cultures that guide educational processes and produce the outcomes of student success. The UK's centuries-long history of higher education has significantly shaped the model that is present today, just as the much shorter history of the US has influenced its educational model. The resulting differences are fodder for debate. Understanding what these differences are and why they exist will help us evaluate whether our current practices and outcomes are adequately preparing students for the societal constraints and challenges engineers will face in the twenty-first century.

References

1. A guide to integrated masters degrees. (n.d.). *Masters Compare*. https://www.masterscompare.co.uk/advice_hub/a-guide-to-integrated-masters-degrees/
2. Alfaa123. (2019). I wanted to get some opinions on a line of thinking that I've held for quite some time. I am NOT an engineer, but I relate [Online forum post]. *Using the title of "Engineer."* Reddit. https://www.reddit.com/r/AskEngineers/comments/csur4f/using_the_title_of_engineer/
3. Arlett, C., Lamb, F., Dales, R., & Hurdle, E. (2010). Meeting the needs of industry: The drivers for change in engineering education. *Engineering Education* 5(2), 18–25. https://doi.org/10.11120/ened.2010.05020018
4. BS in civil engineering. (n.d.) *UTEP*, http://catalog.utep.edu/undergrad/college-of-engineering/civil-engineering/civil-engineering-bs/

5. Bennett, M. (2022, 10 November). Integrated masters degrees—A guide. *The University of Edinburgh.* https://www.findamasters.com/guides/integrated-masters-degrees-guide
6. Bharkhada, B. (n.d.). How has university funding changed?. *MAKEuk.* https://www.makeuk.org/insights/blogs/how-does-university-fundning-affect-the-delivery-of-engineering-degrees
7. Bologna Process. (n.d.). *European University Association.* https://eua.eu/issues/10:bologna-process.html
8. Bridging the skills gap: Career and technical education in high school. (2019, September). *U.S. Department of Education.* https://www2.ed.gov/datastory/cte/index.html
9. Cassidy, Johny. (2023, 16 May). Student loans: How do they work, what can I borrow, and when do I pay it back? *BBC News.* https://www.bbc.com/news/education-62241512
10. Chamorro-Premuzic, T., & Frankiewicz, B. (2019, 19 November). 6 Reasons why higher education needs to be disrupted. *Harvard Business Review.* https://hbr.org/2019/11/6-reasons-why-higher-education-needs-to-be-disrupted
11. Cheston, P. (2012, 13 April). The scandal of Hatfield. *Evening Standard.* https://www.standard.co.uk/hp/front/the-scandal-of-hatfield-7255253.html
12. Cho, R. (2023, 09 February). Climate education in the U.S.: Where it stands, and why it matters. State of the planet. *Columbia Climate School.* https://news.climate.columbia.edu/2023/02/09/climate-education-in-the-u-s-where-it-stands-and-why-it-matters/
13. Churchill, M. (2023, 29 January). The SAT and ACT are less important than you might think. *Inside Higher Ed.* https://www.insidehighered.com/blogs/higher-ed-policy/sat-and-act-are-less-important-you-might-think
14. Cineas, F. (2024, 27 February). Why elite colleges are bringing the SAT back. *Vox.* https://www.vox.com/24083809/college-university-sat-testing-requirement-ivy-league-yale
15. Colby, Glenn. (2023, March). Data snapshot: Tenure and contingency in US higher education. *American Association of University Professors.* https://www.aaup.org/sites/default/files/AAUP%20Data%20Snapshot.pdf
16. Committee on Policy and Advocacy. (2023, January). Employment practices—Use of engineering titles. *National Society of Professional Engineers.* https://www.nspe.org/resources/issues-and-advocacy/professional-policies-and-position-statements/employment-practices-use#:~:text=An%20individual%20whose%20degree%20has,or%20territorial%20engineering%20licensing%20board.
17. Conor. (2022, 19 April). Sandwich courses explained. *Rate My Placement.* https://www.ratemyplacement.co.uk/blog/sandwich-courses-explained/
18. Cost of studying and living in the UK. (n.d.) *British Council.* https://study-uk.britishcouncil.org/moving-uk/cost-studying
19. Countries in the world by population. (2023). *Worldometer.* https://www.worldometers.info/world-population/population-by-country/
20. Daks, M. (2023, 13 March). Engineering a solution to a shortage. *NJBIZ.* https://njbiz.com/engineering-a-solution-to-a-shortage/#:~:text=By%202024%2C%20the%20U.S.%20alone,trying%20to%20plug%20the%20gap.
21. *Department for Education.* (2019, 28 March). Graduate outcomes (LEO): Employment and earnings outcomes of higher education graduates by subject studied and graduate characteristics in 2016/17. *GOV.UK.* https://assets.publishing.service.gov.uk/government/uploads/system/uploads/attachment_data/file/790223/Main_text.pdf
22. Department for Education. (2022, 21 April). Sustainability and climate change: A strategy for the education and children's services systems. *GOV.UK.* https://www.gov.uk/government/publications/sustainability-and-climate-change-strategy/sustainability-and-climate-change-a-strategy-for-the-education-and-childrens-services-systems

23. Edwards, C. (2023, 12 January). Engineering skills crisis: A multi-pronged problem. *Engineering and Technology.* https://eandt.theiet.org/content/articles/2023/01/engineering-skills-crisis-a-multi-pronged-problem/
24. Engineer vs professional engineer: What's the difference? (2019, 18 January). *Encorus Group.* https://encorus.com/2019/01/18/engineer-vs-professional-engineer/
25. Engineer demographics and statistics in the US. (2024, 5 April). *Zippia.* https://www.zippia.com/engineer-jobs/demographics/
26. External Examining Principles. (n.d.). *UK Standing Committee for Quality Assessment.* https://www.qaa.ac.uk/docs/qaa/quality-code/external-examining-principles.pdf?sfvrsn=fe91a281_12#:~:text=External%20examiners%20help%20to%20assess,the%20role%20of%20academic%20staff.
27. Fennell, Andrew. (2023, June). Internship statistics U.S. 2023. *StandOutCV.* https://standout-cv.com/usa/internship-statistics#:~:text=21.5%25%20of%20college%20students%20do,is%20around%201%2C038%20working%20hours.
28. Fernandez, M, & Schwartz, J. (2019, 31 May). Army corps under fire: In a flood, it released more water. *New York Times.* https://www.nytimes.com/2019/05/31/us/army-corps-engineers-midwest-floods.html
29. Field, Kelly. (2022, 22 November). The path to a career could start in middle school. *The Hechinger Report.* https://hechingerreport.org/the-path-to-a-career-could-start-in-middle-school/
30. Garcia, C.A., Weale, S., Swan, L., & Symons, H. (2023, 14 July). Fifth of UK universities' income comes from overseas students, figures show. *The Guardian.* https://www.theguardian.com/education/2023/jul/14/overseas-students-uk-universities-income#:~:text=Fifth%20of%20UK%20universities'%20income%20comes%20from%20overseas%20students%2C%20figures%20show,-Guardian%20analysis%20highlights&text=One%20in%20every%20five%20pounds,tuition%20fees%20for%20financial%20survival.
31. Graddick, S. (2023, 31 January). The US is facing a critical shortage of high tech engineers. *Scripps News.* https://scrippsnews.com/stories/us-facing-critical-shortage-of-high-tech-engineers/#:~:text=According%20to%20the%20U.S.%20Bureau,six%20million%20engineers%2C%20or%20more.&text=In%20the%20heart%20of%20downtown,tech%20companies%20have%20moved%20in.
32. Graduate labour market statistics 2022. (2023, 29 June). *UK.GOV.* https://explore-education-statistics.service.gov.uk/find-statistics/graduate-labour-markets
33. Grey, M. (2019, 11 April). The placement panacea. *Higher Education Policy Institute.* https://www.hepi.ac.uk/2019/04/11/the-placement-panacea/
34. Hatfield rail crash: "Worst example of sustained negligence"—Judge. (2005, 01 November). *New Civil Engineer.* https://www.newcivilengineer.com/archive/hatfield-rail-crash-worst-example-of-sustained-negligence-judge-01-11-2005/
35. Hess, A. J. (2021, 17 August). More than 40% of interns are still unpaid—Here's the history of why that's legal. *Make It.* https://www.cnbc.com/2021/08/17/more-than-40percent-of-interns-are-still-unpaidwhy-thats-legal.html
36. Higher education student data 2021/22. (2023, 31 January). *Higher Education Statistics Agency (HESA).* https://www.hesa.ac.uk/news/31-01-2023/higher-education-student-data-202122
37. Holmes, B., & Duda, C. (2020, 14 December). Applying to college as undecided major: Pros, cons. *U.S. News and World Report.* https://www.usnews.com/education/blogs/college-admissions-playbook/articles/pros-cons-of-applying-to-college-as-an-undecided-major

38. Hurix. (2023, 22 February). Top 6 teaching strategies adopted by higher-ed institutions post-Covid. *Hurixdigital*. https://www.hurix.com/teaching-strategies-adopted-by-higher-ed-institutions-post-Covid/
39. Hylton, P., & Otoupal-Hylton, W. (2016). Comparison of engineering education in the United States Versus the United Kingdom. *2016 ASEE Annual Conference & Exposition*. https://peer.asee.org/comparison-of-engineering-education-in-the-united-states-versus-the-united-kingdom.pdf
40. Johnson, R. (2023, 20 March). New survey finds most college grads would change majors. *BestColleges*. https://www.bestcolleges.com/blog/college-graduate-majors-survey/
41. Kaplan, Z. (2023, 17 April). 20+ Internship statistics students need to know. *Forage*. https://www.theforage.com/blog/basics/internship-statistics#:~:text=Many%20Students%20Intern%2C%20but%20Even,Want%20To%20%E2%80%94%20And%20Can't&text=However%2C%20new%20data%20from%20the,%25)%20had%20internships%20in%202021.
42. Knox, L. (2023, 26 April). "We're not slowing down," student workers say. *Inside Higher Ed*. https://www.insidehighered.com/news/faculty-issues/labor-unionization/2023/04/26/were-not-slowing-down-student-workers-say
43. Lafley, A.G. (2012, 05 February). A liberal education: Preparation for career success. *Huffpost*. https://www.huffpost.com/entry/a-liberal-education-prepa_b_1132511
44. London's geography and population. (n.d.). *Trust for London*. https://trustforlondon.org.uk/data/geography-population/
45. MPs and their degrees: Here's where and what our UK politicians studied. (2019, 13 December). *Studee*. https://studee.com/media/mps-and-their-degrees-media/
46. Making the grade? How state public school science standards address climate change. (2020, October). *National Center for Science Education and the Texas Freedom Network Education Fund*. https://ncse.ngo/files/MakingTheGrade_Final_10.8.2020.pdf
47. Maples, B. (2021, 02 August). What is a student union? *University Compare*. https://universitycompare.com/advice/student/student-union
48. McCammon, E. (2022, 31 October). SAT/ACT Prep Online Guides and Tips. *PrepScholar*. https://blog.prepscholar.com/minimum-sat-score-for-college#:~:text=Less%20selective%20public%20institutions%2C%20as,policies%20and%20your%20other%20qualifications
49. McDonald, J. (2021, 24 June). Engineering zero: Bringing the engineering profession together to meet the net zero ambition. *Royal Academy of Engineers*. https://raeng.org.uk/blogs/engineering-zero-bringing-the-engineering-profession-together-to-meet-the-net-zero-ambition
50. Muniz, H. (2023, 21 March). What is a Good ACT Score? *BestColleges*. https://www.bestcolleges.com/blog/what-is-a-good-act-score/#:~:text=Schools%20vary%20considerably%20in%20the,in%20the%2032%2D36%20range
51. NASA. (2023, 09 June). Do scientists agree on climate change? *Global Climate Change: Vital Signs of the Planet*. https://climate.nasa.gov/faq/17/do-scientists-agree-on-climate-change/#:~:text=Yes%2C%20the%20vast%20majority%20of,global%20warming%20and%20climate%20change
52. NUS Charity. (n.d.) Our impact on history. *NUS UK*. https://www.nusconnect.org.uk/nus-uk/who-we-are/our-story/our-impact-on-history
53. NUS UK. (2022, 11 June). NUS comment on student loan interest rate cap. NUS UK. https://www.nus.org.uk/nus_comment_on_student_loan_interest_rate_cap
54. NUS UK. (2022, 08 November). 70,000 university staff across the UK to strike on 24th, 25th and 30th November. *NUS UK*. https://www.nus.org.uk/nus_uk_response_to_ucu_strikes_announcement

References

55. National Academy of Engineering. (2004). The engineer of 2020: Visions of engineering in the new century. *National Academy of Sciences*. National Academies Press. https://nap.nationalacademies.org/download/10999
56. National Center for Education Statistics. (2023, May). Undergraduate enrollment. *Institute of Educational Sciences*. https://nces.ed.gov/programs/coe/indicator/cha
57. Neitzel, M. T. (2023, 13 June). The test-optional college admissions movement continues to grow. *Forbes*. https://www.forbes.com/sites/michaeltnietzel/2023/06/13/the-test-optional-college-admissions-movement-continues-to-grow/?sh=14a09ed81326
58. New survey: student confusion selecting majors increases higher education cost and time to earn a degree. (2019, 14 October). *Businesswire*. https://www.businesswire.com/news/home/20191014005009/en/New-Survey-Student-Confusion-Selecting-Majors-Increases-Higher-Education-Cost-and-Time-to-Earn-Degree
59. Parker, A. (2023, 28 February). International students in the US. *Prosperity for All*. https://www.prosperityforamerica.org/international-students-in-the-us/#:~:text=during%20that%20time.-,Number%20Of%20International%20Students%20In%20The%20U.S,the%20U.S%20in%202020%2D21
60. Porter, I. (2024, 14 March). Why some top colleges are requiring the SAT again. *Christian Science Monitor*. https://www.csmonitor.com/USA/Education/2024/0314/Why-some-top-colleges-are-requiring-the-SAT-again#:~:text=Yale%2C%20Brown%2C%20and%20Dartmouth%20are,optional%2C%20citing%20the%20same%20reason
61. Professional standards framework (PSF 2023). (2023). *AdvanceHE*. https://www.advance-he.ac.uk/teaching-and-learning/psf
62. Protecting students as consumers. (2023, 15 June). *Office for Students*. https://www.officeforstudents.org.uk/publications/protecting-students-as-consumers/#:~:text=While%20the%20idea%20of%20students,should%20contain%20fair%20terms%20and
63. Reward and recognition of teaching in higher education. (2009, February). *The Higher Education Academy*. https://s3.eu-west-2.amazonaws.com/assets.creode.advancehe-document-manager/documents/hea/private/reward_and_recognition_interim_2_1568037076.pdf
64. Root cause analysis of the Hyatt Regency disaster—Cautionary tale about assumptions. (2023). *ThinkReliability*. https://www.thinkreliability.com/case_studies/root-cause-analysis-of-the-hyatt-regency-disaster-cautionary-tale-about-assumptions/
65. Rose, S. (2020). The importance of having direction in life. *Steve Rose PhD Counselling*. https://steverosephd.com/the-importance-of-having-direction-in-life/#:~:text=Also%2C%20having%20a%20sense%20of,adherence%20to%20long%2Dterm%20goals
66. Royal Academy of Engineering. (2007, June). Educating engineers for the 21st century. *Royal Academy of Engineering*. https://raeng.org.uk/media/rdjje5xo/educating_engineers_21st_century.pdf
67. Schaeffer, K. Nearly all members of the 118th Congress have a bachelor's degree—and most have a graduate degree, too. (2023, 02 February). *Pew Research Center*. https://www.pewresearch.org/short-reads/2023/02/02/nearly-all-members-of-the-118th-congress-have-a-bachelors-degree-and-most-have-a-graduate-degree-too/
68. "Series of errors" Led to Hatfield crash. (2021, 8 May). *BBC News*. http://news.bbc.co.uk/1/hi/uk/1318273.stm
69. Smurthwaite, J. (2023, 28 April). Understanding the undergraduate grading system in the UK. *Hotcoursesabroad*. https://www.hotcoursesabroad.com/study-in-the-uk/applying-to-university/understanding-undergraduate-grading-system-in-uk/
70. Statement of changes: UK immigration rules announced. (2023, 19 July). *Morgan Lewis*. https://www.morganlewis.com/pubs/2023/07/statement-of-changes-uk-immigration-rules-announced#:~:text=From%201%20January%202024%2C%20international,are%20eligible%20to%20do%20so

71. Status of engineers. (n.d.) *Engineering Council*. https://www.engc.org.uk/glossary-faqs/frequently-asked-questions/status-of-engineers/
72. Sun, K-L. (2022, 22 August). What does it mean to be undeclared? *BestColleges*. https://www.bestcolleges.com/blog/what-does-undeclared-mean/#:~:text=An%20estimated%2020%2D50%25%20of,once%20in%20their%20college%20career
73. Taylor, A., Nelson, J., O'Donnell, S., Davies, E., & Hillary, J. (2022, 03 March). The skills imperative 2035: What does the literature tell us about essential skills most needed for work? *National Foundation for Educational Research*. https://www.nfer.ac.uk/the-skills-imperative-2035-what-does-the-literature-tell-us-about-essential-skills-most-needed-for-work/
74. Teaching and learning accreditation. (n.d.) *AdvanceHE*. https://www.advance-he.ac.uk/membership/teaching-and-learning-accreditation
75. Tenure. (n.d.) *American Association of University Professors*. https://www.aaup.org/issues/tenure#:~:text=The%20principal%20purpose%20of%20tenure,conduct%20research%20in%20higher%20education
76. The UK quality code for higher education. (2023, May). *Quality Assurance Agency for Higher Education*. https://www.qaa.ac.uk/docs/qaa/quality-code/revised-uk-quality-code-for-higher-education.pdf?sfvrsn=4c19f781_24
77. The Sutton Trust. (2017, 10 June). Parliamentary privilege—The MPS 2017. *The Sutton Trust*. https://www.suttontrust.com/our-research/parliamentary-privilege-the-mps-2017-education-background/
78. The UK standard for professional engineering competence and commitment (UK-SPEC). (2020, August). 4th ed. *Engineering Council*. https://www.engc.org.uk/media/4338/uk-spec-v14-updated-hierarchy-and-rfr-june-2023.pdf
79. Tyson, A., Funk, C., & Kennedy, B. (2023, 18 April). What the data says about Americans' views of climate change. *Pew Research Center*. https://www.pewresearch.org/short-reads/2023/04/18/for-earth-day-key-facts-about-americans-views-of-climate-change-and-renewable-energy/#:~:text=Climate%20change%20is%20a%20lower,and%20reducing%20health%20care%20costs
80. U.S. Department of Labor. (2018, January). Fact sheet #71: Internship programs under the Fair Labor Standards Act. *U.S. Department of Labor*. https://www.dol.gov/agencies/whd/fact-sheets/71-flsa-internships
81. UK higher education providers—Advice on consumer protection law. (2023, 31 May). *Competition and Markets Authority*. https://assets.publishing.service.gov.uk/government/uploads/system/uploads/attachment_data/file/1159885/Consumer_law_advice_for_higher_education_providers_.pdf
82. UK vs US education system. (n.d.). International Student. https://www.internationalstudent.com/study-abroad/guide/uk-usa-education-system/
83. United States population. (2023, 16 July). Worldometer. https://www.worldometers.info/world-population/us-population/#:~:text=the%20United%20States%20population%20is,96%20people%20per%20mi2).&text=The%20median%20age%20in%20the%20United%20States%20is%2038.1%20years
84. Universities UK. (2022, 20 Dec.) International facts and figures 2022. *UUKI Publications*. https://www.universitiesuk.ac.uk/universities-uk-international/insights-and-publications/uuki-publications/international-facts-and-figures-2022#:~:text=In%202020%2D21%2C%20the%20UK,student%20population%20in%202020%2D21
85. Vyas, K. (2023, 22 March). 23 of the worst engineering disasters to date. *Interesting Engineering*. https://interestingengineering.com/lists/23-engineering-disasters-of-all-time

86. Weale, S. (2023, 31 May). Funding model for UK higher education is 'broken', say university VCs. *The Guardian*. https://www.theguardian.com/education/2023/may/31/funding-model-for-uk-higher-education-is-broken-say-university-vcs
87. What is the ACT? (2023). *The Princeton Review*. htttps://www.princetonreview.com/college/act-information
88. What is the SAT? (2023). *The Princeton Review*. https://www.princetonreview.com/college/sat-information
89. What is the student's union and how can they help me? (2023, 21 July). *GBMag*. https://greatbritishmag.co.uk/student-guide/what-is-the-students-union-and-how-can-they-help-me/
90. Whitford, E., with Howard, C., eds. (n.d.) America's top colleges. *Forbes*. https://www.forbes.com/top-colleges/
91. Young, S. (2017, 28 February). The four-year option: Why take an integrated master's? *The Guardian*. https://www.theguardian.com/education/2017/feb/28/the-four-year-option-why-take-an-integrated-masters

Preparing Engineering Students for Leadership

Engineers as Leaders

Engineering has always been a complex combination of technology, materials, and human factors. That complexity is revealed in the catastrophic failures of the *Titanic* and the Space Shuttles *Challenger* and *Columbia*. Most engineers are familiar with these tragedies, but many are less aware that a lack of professional skills such as leadership, communication, and other human factors played a major role in causing these disasters.

The *Titanic* was hailed as a unique and fantastic feat of engineering, but four days after setting sail, it had transformed into a catastrophic failure. The *Titanic* disaster has resulted in positive changes in design, materials, risk analysis and prevention, communication, and other factors that have improved maritime safety, but those changes came at the cost of over 1500 lives.

On the morning of April 14, 1912, Edward Smith, *Titanic's* captain, was warned about nearby icebergs. But the White Star Line that owned *Titanic* was pressuring him to reach New York early, in time for a celebration that would generate publicity and revenue for the line. Despite known risk, he kept the ship speeding along at about 22 knots [38, 51].

Secure in the belief that "God himself could not sink this ship," the White Star Line had equipped *Titanic* with only 16 lifeboats, enough for about one fourth of the passengers on board. Engineers believed the ship's double-bottomed hull and 15 transverse bulkheads made *Titanic* impervious to icebergs. It was designed to stay afloat even if four "watertight" compartments were flooded. But when the 53,310-ton ship collided with a 500,000-ton iceberg, multiple flaws were revealed [2].

Five compartments filled with water. The more than three million rivets holding *Titanic's* hull together were made of poor-quality iron with a high slag content, making them brittle and less ductile, especially in the North Atlantic's freezing temperatures.

"*Titanic*'s collision with the iceberg caused the rivet heads to break off, popped the fasteners from their holes and allowed water to rush in between the separated hull plates" [41].

Design and materials failures were compounded by human and technological failures. Collapsible lifeboats had been stowed out of sight, and the crew did not know how to assemble them. No lifeboat drill had been done at sea. Below decks, 670 immigrants in third class (steerage) were trapped behind locked doors by order of the US Immigration Department [2].

Titanic's Marconi radio system "had long since been superseded by other radio pioneers ... Marconi used his patents, research and monopoly power to hold back competition from other systems." Any message sent through Marconi's "radio telegraph ... soaked up virtually all of the frequency bandwidth," interrupting the signal of messages from other ships [31].

Approximately two hours before colliding with the iceberg, *Titanic's* Senior Wireless Officer Jack Phillips received a message from the steamship *Mesaba*, warning of multiple icebergs and an icefield directly in *Titanic*'s path, but he did not relay the message to the bridge. Phillips was busy sending and receiving personal messages for passengers—messages that generated high profits for the Marconi company. Although Phillips was under the command of *Titanic's* captain, he was paid by and accountable to Marconi [17].

About an hour later, Phillips received a second warning from Cyril F. Evans, a radio operator on the *Californian*. The message warned that the *Californian* was "stopped and surrounded by ice." Evans later told a US Senate inquiry that Phillips messaged back, "'Shut up, shut up, I am busy ...'" [31]. Since Phillips made no further response, Evans took off his headphones and went to sleep. There was no further communication with the *Californian*, which was only 20 miles away from *Titanic* and could have arrived in time to rescue passengers.

At about 1:40 a.m. on April 15, *Titanic* sent this message: "SOS SOS CQD [Come Quick Distress] CQD—MGY [*Titanic*'s call sign] We are sinking fast passengers being put into boats" [33]. But it was too late. At 2:20 a.m., over 1500 passengers lost their lives as *Titanic* sank beneath the waves.

Engineers have learned a great deal since 1912, yet we continue to grapple with problems that mirror the root causes of the *Titanic* disaster. In 1986, the Space Shuttle *Challenger* exploded 73 seconds after launch when an O-ring failed, allowing "pressurized burning gases to break through a joint, causing one of the booster rockets to spin out of control and the orbiter to break apart." Investigations revealed that NASA knew the O-ring could fail at temperatures lower than 53° Fahrenheit, yet the flight readiness documents contained no mention of this risk [18].

Morton Thiokol, the company that built *Challenger*'s booster rockets, sent engineer Allan McDonald and his team to NASA to approve the launch, a required safety protocol. But McDonald's team refused to sign off, citing three high-risk factors including the 18-degree weather forecast. "McDonald later recounted that after some pressure from NASA, the engineers' concerns were overridden, and his boss instead signed off on the launch." NASA was relying on the *Challenger* flight for a much-needed PR boost. Schoolteacher Christa McAuliffe, the first civilian to participate in a shuttle crew, was planning to broadcast a lesson live during orbit, and President Reagan planned to highlight McAuliffe in his upcoming State of the Union Address. Any delay would have interrupted both of those plans. PR won over safety, and millions of schoolchildren watched as *Challenger* exploded, killing all seven crew members. The Rogers Commission concluded that a root cause of the disaster was a "serious flaw in the decision-making process leading up to the launch" [18].

Another disaster occurred in 2003. As Space Shuttle *Columbia* was launched on its 28th flight, a piece of foam fell from a structure connecting an external tank to the spacecraft. "Several people within NASA pushed to get pictures of the breached wing in orbit. The Department of Defense was reportedly prepared to use its orbital spy cameras to get a closer look. However, NASA officials in charge declined the offer." Columbia exploded during re-entry, killing its seven-member crew. An investigation determined that the foam had struck the left wing, leaving a hole that leaked gases from the atmosphere into the shuttle during re-entry, causing the explosion. We will never know if NASA could have taken steps to prevent the explosion, but investigators found that NASA had known about the problem with the foam for years and had done nothing to correct the issue [25].

Professional Development

These disasters from 1912, 1986, and 2003 reveal a broad spectrum of ongoing problems rooted in a combined systemic failure of technology, expertise, materials, and human factors. Historically, engineering education has focused on the technical expertise necessary to design and manufacture a working product. The *Titanic* and more recent failures reveal that technical expertise is essential, but it is not sufficient. Engineers must be prepared to lead society in solving some of its most challenging issues. They must be able to understand and anticipate the logistical, social, ethical, economic, and political impacts of their designs and decisions. They must also be able to interact and build professional relationships across a wide variety of professional groups, recognizing and valuing contributions from people outside the technical fields. Engineers' decisions and actions directly impact people's lives. As Grimson explains,

> Engineering lies at the interface between science on the one hand and society on the other. It is concerned with the systematic application of scientific and mathematical principles towards practical ends *for the benefit of people* [emphasis added]. Traditionally the emphasis in engineering education has been on the scientific side, with students given a thorough grounding in the basic scientific and mathematical principles underpinning their discipline. However, the constraints on engineering problem-solving today are increasingly not technical, but rather lie on the *societal and human side of engineering practice* [emphasis added]. (Grimson [22, p. 31])

That *"societal and human side"* is an essential part of an engineering education. Engineers must understand and work in concert with society both locally and globally. Human beings across industries and continents impact each other now more than ever before. Our job as educators is to prepare students for the technical—as well as the human and societal—aspects of engineering. We are proud of our profession, so we tend to think we are doing a good job. But maybe we are less successful than we would like to think. Are we providing students with an education that prepares them to work in a global, technical, societal, and human environment?

Alarmingly, industry—one of our primary customers—thinks we are failing. Koromyslova et al. report on industry perceptions:

> In the last 20 years, employers have expressed concerns about the preparation of their newest engineering professionals. While they are usually very satisfied with their new engineer's technical preparation, they are dissatisfied with the ability of those new hires to function effectively in a professional environment. Professional skills have long been listed as some of the most important skills necessary for a successful engineering career, yet there was little change in academia toward improving graduate professional skills. (Koromyslova et al. [30], p. 2)

Due to that "little change in academia," the problem is worsening:

> During the last decade, the National Association of Colleges and Employers has been reporting that less than 50% of employers evaluate college graduates as proficient in competencies such as professionalism/work ethic, oral/written communications, teamwork/collaboration, leadership, and other related skills. (Koromyslova et al. [30, p. 2])

Students, however, have a very different perception: "a majority of college graduates (over 80%) believe they are proficient in these skills when entering their first workplace." There is also a significant gap between the perceptions of academic department heads and employers: "researchers discovered that while 52% of department heads considered skills to be strong, only 9% of employers considered these same skills as strong for graduates of engineering professional programs" [30, pp. 1, 5].

Other researchers find similar gaps. Daley and Baruah highlight a "disconnect between the *leadership* [emphasis added] needs of industry and what undergraduate courses attribute the most importance to" [10, p. 25]. Chan et al. agree that "engineers need to possess a set of personal and professional skills *including leadership skills* [emphasis added] in addition to their expected technical competencies" [6, p. 1251].

Calls for including professional skills—including leadership skills—in engineering education date even further back than the 2004 publication of *The engineer of 2020: Visions of engineering in the new century* [39]. Daley and Baruah explain why engineers need leadership skills: "the modern global economy has rendered technical skills alone insufficient: communication, project management, and *other leadership skills* [emphasis added] are becoming more critical than ever before" [10, pp. 1–2].

Rottman et al. note that over the last two decades, accreditation boards and researchers have increasingly called for reforming engineering education to include leadership skills. Like Daley and Baruah, they point to the global economy: "engineers who supplement their technical training with leadership education will be well positioned to compete in the increasingly global market" [47, p. 147]. Conversely, "if graduates of engineering programs are exclusively trained in technical problem solving," they will "fall behind other nations" [48, p. 3]. Currently, China and India supply about 60% of the world's engineers. *Inside Higher Ed* published the following numbers of 2023 engineering graduates: China—600,000, India—350,000, America—70,000 [15].[1] In addition to greater numbers, engineers from China and India can be hired for about 30% of the cost of hiring an engineer from the UK or the US. The leadership skills demanded by industry are essential for UK and US engineers to maintain a competitive advantage in the global marketplace. Daley and Baruah assert, "Researchers have concluded that engineers across sectors do have the potential to lead and there is now a need of a widespread recognition of engineering as a leadership profession" [10, p. 25].

Rethinking and redesigning our engineering programs and curricula to meet student needs and industry demands is no easy task. One of the initial barriers that must be overcome is academic and faculty resistance to applying the term *leadership* to engineering. Kendall et al. found that "a critical mass of engineers continue to resist the notion that engineering is a leadership profession" [27]. Similarly, research by Rottman et al. revealed that "many engineers resisted the idea of leadership because they found it to be inconsistent with their professional identities as engineers." In fact, most felt that "'leadership' was a suspect term ... imprecise, impractical, elitist, and just 'not us'" [48, pp. 7, 8]. Some resistance is based on the assumption that *leadership* is a positional term. But engineering leadership (EL) is not necessarily positional (indicating executive or managerial roles); instead, it is relational: "engineering leadership as a concept can embody notions of influence, personal effectiveness, collaboration, and engineering competency" [6, p. 1251].

Rottman et al. identify four reasons for the perceived dissonance between engineering and leadership:

- "For individuals whose love of engineering comes from their technical problem solving, the sudden shift to resolving 'people problems' can feel both uncomfortable and un-engineer-like."
- "Engineers who hold a traditional, hierarchical view of leadership may experience the phenomenon as inconsistent with the egalitarian, team-based norms of their discipline."
- "The somewhat amorphous term 'leadership' does not always resonate with members of an occupational group whose reputations hinge on their technical precision."

---

[1] The number of UK engineering graduates was not included in this report. However, other reports place UK numbers significantly below US numbers, as would be expected based on size and population.

- "When faculties of engineering support leadership primarily through optional, extracurricular involvement, ... a critical mass of students may view it as peripheral to the core curriculum." [48, pp. 1–2]

Polmear et al. [44, p. 953] note that "For most engineering faculty, leadership may be seen as outside their expertise, which can create a reticence to integrate it into their classroom." My own research confirms this. As part of my year-long faculty research fellowship, I conducted over 200 interviews with academics from nearly 30 institutions in the UK. Also, in my role as an engineering program evaluator for the Accrediting Board for Engineering and Technology (ABET), I have interviewed engineering faculty in universities across the US. On both sides of the Atlantic, academics and faculty have often told me that they do not feel competent to take on the task of teaching professional skills, especially leadership. A large percentage of academics and faculty progressed from graduate school straight into teaching and have little if any experience in the business, social, and political interactions of industrial or government engineering. Many faculty have spent their entire careers in academia. Their own professional skills have developed in a narrow academic world that is not consistent with the needs and practices of industry and culture. University departments—and even individual academics within departments—are often siloed. For example, if an academic teaches only dynamics, they can easily lose touch with other academics within the department rather than sharing with and learning from them. They are likely to have very little, if any, interaction with academics from other departments in the university. Academics often teach and research in a single area of specialization and are rewarded if they achieve excellence in that specialization, which further decreases motivation to learn from and collaborate with others. How ironic is it that we tell students they will need to work on teams with people from multiple disciplines, including those outside engineering, but our work drives us to do exactly the opposite of what we teach students to do.

Yet the concerns of faculty are real. Many feel highly competent to teach engineering but completely unprepared to teach leadership. Polmear et al. note that although faculty may be untrained in leadership skills, "they will nevertheless need to help their students develop them." They also found that engineering educators have often received more leadership training and modeling than they realize. They can be guided to identify "leadership opportunities in their current role, industry experience, personal experiences and relationships ... as impactful in shaping how they learned about leadership." Polmear et al. advise that along with guiding academics to reflect on experiences that have contributed to developing their own insights and skills, universities should provide formal professional training. That training should be accompanied by "hiring individuals with different backgrounds, and building a culture and curriculum supportive of leadership education" [44, pp. 963–964].

If a department creates a leadership-infused curriculum but does not simultaneously create a leadership culture, internal resistance from colleagues can become a barrier for faculty working to integrate leadership, especially for younger or newer faculty who do not have the "capital" to implement changes and overcome such resistance. In the absence of a departmental leadership culture, few faculty independently attempt to integrate leadership training in their pedagogy, and those who do are often treated as outcasts by other faculty. Leadership culture is also critical for students who may otherwise feel that leadership is peripheral to engineering [44, p. 962–965].

The call for engineering leadership is loud and strong, and some universities in both the UK and in North America have responded with varied integrations of leadership training. One of the earliest engineering leadership programs was developed at Tufts University in 1987. However, leadership integration did not become prominent until the early 2000s [29]. Since then, leadership in varying degrees has been added to engineering programs on both sides of the Atlantic. Daley and Baruah researched the prevalence of leadership skills in 19 of the 20 UK Russell Group universities that offer a BEng degree. While noting that "the majority of the evidence of leadership integration in engineering curricula seems to be in the USA rather than Europe," they highlight Loughborough University and The University of Bristol (both of which I visited as part of my research fellowship) along with US universities Massachusetts Institute of Technology (MIT) and The University of Texas at El Paso (UTEP) (where I serve as the Founding Director of the Engineering Innovation and Leadership program for the College of Engineering) as having "integrated programmes which explicitly develop leadership skills" [10, p. 6].

These integrations are making progress toward preparing students to tackle current and future engineering challenges, but we still have much to learn. Daley and Baruah note that in the module specifications of BEng courses at the 19 Russell Group universities they researched, the most frequently mentioned themes were analysis, problem solving, written communication, explanation, teamwork, and project management. The least frequently mentioned were study of leadership, influencing others, entrepreneurship, marketing, risk management, change management, and peer evaluation. Daley and Baruah highlight internal disconnects within these programs, such as the fact that 15% of modules referenced teamwork, but only 5% referenced peer evaluation. Only 3.5% referenced both teamwork and peer evaluation. Importantly, some of the skills that received the least attention are those most in demand by industry [10, pp. 13–14, 21].

Polmear et al. found that in the US, although ABET requires students to learn leadership skills, several engineering faculty they interviewed openly admitted that "they were not doing anything specific to teach leadership to their students as they felt unprepared to do so." Others said their attempts were limited to modeling leadership or recommending that students participate in extracurricular activities [44, pp. 953, 962–963].

Daley and Baruah make the important point that to successfully include leadership training in engineering education, "There is a need for educators to first understand these skills and then design curricula and teaching approaches that align with the needs of the industry" [10, p. 21]. Implementing leadership in engineering education will require us to face two challenges:

1. Definition: How will we define engineering leadership? What knowledge, skills, and experiences will it include?
2. Inclusion: What are the best methods for including leadership in existing engineering programs?

Definition
Polmear et al. recognize the complexity of defining *leadership:*

> In the United States of America (USA), there is agreement across industry ... and education ... regarding the importance of leadership, but there is little agreement about its definition in engineering. Leadership has been conceptualized as a process that involves influence, a group and a common goal. Within the engineering context, leadership is often associated with a vertical approach that positions an individual leader in charge of followers. ... Other definitions in engineering are structured around the competencies that enable leadership. From the perspectives of engineering undergraduates and professionals in one study, leadership involves intrapersonal attributes, impacts of engineering on society, interpersonal relationships, and abilities to adapt to change. ... Leadership has also been defined as a set of 19 interpersonal, intrapersonal, and technical competencies such as communication, critical thinking, ambition, humility, and management .... Looking across these definitions reveals *the importance of situating leadership in the engineering context* [emphasis added] and developing the requisite competencies to promote its development. (Polmear et al. [44, pp. 950–951])

Many scholars have offered definitions of engineering leadership, but there is no consensus. In fact, a variety of labels other than leadership are used to describe the professional skills included in leadership. Khalid et al. [28] use the term *soft skills*. Riemer [45] describes these skills as *EQ* (Emotional Intelligence). The Engineering Ethics Reference Group [13], empaneled by the Royal Academy of Engineering (RAE) and the Engineering Council, defines the needed professional skills as one of five themes under the heading *Ethics*. In the US, the Accreditation Board for Engineering and Technology (ABET) defines essential professional skills as *Student Outcomes* that "prepare students to enter the professional practice of engineering." One of those outcomes is stated as "an ability to function effectively on a team whose members together *provide leadership* [emphasis added], create a collaborative and inclusive environment, establish goals, plan tasks, and meet objectives" [1].

Daley and Baruah studied 19 BEng courses at Russell Group universities in the UK. Their report illustrates the complexity of definition: "there is arguably little or no consensus on an agreed definition of leadership." One source cited in their research identified 221 definitions from 587 publications; another discovered 90 different variables. After extensive research, Daley and Baruah developed a list of the "knowledge bases, or skill areas" that comprise their comprehensive definition of *leadership* for engineers:

- Character development
  - Environmental considerations
  - Ethics
  - Professional conduct
  - Societal impact of engineering
- Business knowledge
  - Accounting and finance
  - Business strategies
  - Economics
  - Law
  - Marketing
  - Entrepreneurship
- Interpersonal skills
  - Analysis skills
  - Innovation
  - Problem-solving skills
  - Self-evaluation
- Management
  - Change management
  - Project management
  - Resource management
  - Time management
  - Risk management [10, pp. 7–9].

Most—if not all—engineering education programs in the UK and the US have already-crowded curriculums. In the US, universities have been required to reduce the number of credit hours outside of the core curriculum. In the UK, students must be able to complete the BEng in just three years. These constraints have made it more difficult to accomplish curricular and learning outcomes required for accreditation. In this context, incorporating elements like Daley & Baruah's four main topics and 19 subtopics may sound impractical or even impossible.

Kendall et al. help by pointing to the key element of definition—value:

> Ultimately, what we describe as engineering leadership is a *statement of what we value* [emphasis in original]. Together with others, we argue that engineering leadership is not merely technical, nor is it merely managerial. It includes leadership at the intersection of the two: the social impact of technological innovations. (Kendall et al. [27, p. 17])

This definition of engineering leadership as "the social impact of technological innovations" explains how engineering modifies broader definitions of leadership and aligns with the call from Kendall et al. for an "engineering-specific approach to student leadership development." They offer the following definition that includes "three key elements that express the values of our field":

Engineering Leaders (a) employ the full range of engineering skills and knowledge in the design of socio-technical innovations, while (b) seeking to understand, embrace, and address the current and future impact of their work in context by (c) actively fostering engaged and productive relationships with diverse stakeholders, including themselves and their team, the users of their technologies, and those impacted by their engineering work. (Kendall et al. [27, p. 17])

The authors also provide a rationale for their definition of engineering leadership:

Because these three practices are inextricably linked, they enable engineers to consider the larger context of their work from the outset, not retroactively. Furthermore, these are behaviours that engineers will continue to develop throughout their professional careers. Accordingly, these behaviours must be integrated from the very start of an engineer's training in the field to provide a foundation for practicing engineers to build on. (Kendall et al. [27 p. 17])

Rottman and Kendall explain how value informs program-specific definitions of engineering leadership:

When EL [Engineering Leadership] program developers get together, we tend to discuss a few key topics related to program design, pedagogy, program assessment, securing funding and justifying programs to university administrators. While these tactical considerations demand our ongoing attention, they can be difficult to address without first examining what we hope to accomplish. Stated differently, how may we answer the question—leadership education for what? This goes beyond rationalizing the existence of our respective programs to senior administrators or funders. It is a more fundamental philosophical question about the nature of knowledge, learning, and intended impact. (Rottman and Kendall [46, pp. 149–150])

Reforming curriculum to include leadership does not require us to adopt or teach a specific theory of leadership. Engineering leadership is not about theory; it is about outcomes and practice: learning how to inspire change, think in new ways, work with subject matter experts, inspire and value contributions from team members—including and perhaps especially those from fields outside of engineering—and listen to people who will be impacted by our work. It includes learning to take the initiative to identify problems and how and where to seek out solutions. We know that as engineers gain workplace experience, most will develop at least some leadership skills, but that development may take 5, 10, or even 15 years of a 30-year career. We must jumpstart and accelerate the process so that students are better prepared for the expectations and needs of industry at the beginning of an engineering career. The world is facing unprecedented challenges. If engineers are to solve the world's problems now and in the future, we must graduate engineers who are not only technically competent but also possess the *foundational* leadership skills that will facilitate engineering practice in the modern world.

Inclusion into the Academic Curriculum
Including leadership development in our university programs also requires that we overcome a second critical barrier: finding the appropriate method of inclusion. Much more research needs to be done on the best ways to accomplish this inclusion, but three methods have been suggested.

Standalone Course(s)

The simplest approach is a standalone course, often in partnership with other university departments (e.g., humanities, business, and communication). Daley and Baruah note that among the Russell Group universities they studied, professional development skills have been taught through both "optional and compulsory modules, with slightly greater occurrence in compulsory modules" [10, p. 25].

While standalone courses may provide a solution in some cases, this method is not without its problems. One is that standalone courses must often be presented as electives, so students may choose not to take them. Another problem is that when leadership skills are separated from engineering coursework, students do not learn that engineering and leadership coexist. Polmear et al., found that relegating professional skills to courses outside of engineering sends a message to students that these subjects are tangential or unnecessary for their engineering career:

> Beyond what is covered in the formal curriculum, students also learn through what is not taught, termed the null curriculum, and what is implicitly transmitted, called the hidden curriculum. The lack of curricular coverage of a topic sends a message to students about its importance and the privileging of technical content implies what is valued in the profession. These aspects of the hidden curriculum can thus influence students' understanding of leadership in engineering. (Polmear et al. [44, p. 966])

Polmear's research indicates that students tend to resist when they "perceive leadership as tangential to engineering when it is taught in standalone courses or modules or is not explicitly taught in their engineering coursework" [44, p. 965]. Thus, relying on standalone courses has limited value toward achieving engineering leadership goals.

Placements, Co-curricular Activities, and Active Learning

A second option for inclusion is to rely on professional placements and co-curricular activities along with active learning strategies. Placements usually occur after the second or fourth year and last between 12 and 15 months. Some university placement programs choose to work with specific employers that collaborate with the university department to assure that students learn the necessary skills. Another strategy is to encourage students to participate in co-curricular activities such as "professional and honor societies, clubs, community outreach, networking events, study abroad, internships or cooperative education opportunities, extracurricular engineering design competitions, and other related activities outside the curriculum." While none of these things provide formal, intentional instruction in professional skills, they do provide the opportunity to practice and improve professional skills" [30, p. 2], though often through trial and error and without reflection.

Active learning strategies such as in-class team and group projects are also cited as opportunities for students to develop leadership competencies such as collaborating with others to achieve goals. Collaboration is an essential part of leadership, but it does not equal or replace leadership. A leader stimulates others to go beyond current knowledge, work toward developing a process that may seem impossible, or take on a problem that has no clear answer. Without leadership, a collaborative group is more likely to work within what is already known. Leadership is less about learning to do something correctly and more about catalyzing learning so that a previously unforeseen or unimagined solution can be discovered.

Active learning opportunities are important, but some researchers question whether providing opportunities is sufficient. "Although engineering programs require group work/projects, it is often assumed that students will learn teamwork skills naturally from the group assignments" [30, p. 11]. Without prior instruction and reflection, students may fail to learn the necessary skills and practices. Thus, as Daley and Baruah point out, *"Currently, engineers hone their leadership and management skills while at work (i.e. learning 'soft skills the hard way')"* [emphasis in original] [10, p. 4]. Certainly, students should continue to learn even after they graduate, but to meet industry demands for professional and leadership skills, students' training must begin as early as possible in their academic careers, thus *accelerating the leadership effectiveness* in their professional careers.

Curricular Integration
The option most frequently recommended by researchers and leadership practitioners is to integrate professional skills development into engineering module curriculum. Doing so requires significant work and a long-term plan and often includes changing the culture within an engineering department. Truly successful integration also requires a radical shift in curriculum and methodology. McDonald and Jamieson surveyed seven engineering leadership programs that have progressed to the point of identifying specific competencies and grown "beyond a single champion." They summarize common elements of these integrations in Fig. 1.

Daley and Baruah note that successful leadership integrations illustrate "the power of blending education with experiential learning and reflection, alongside opportunities to facilitate leadership capabilities and skills" [10, p. 6]. However, as previously noted, an already-crowded curriculum makes this blending a challenging task. In 2007, as we

Fig. 1 Engineering leadership definition synthesis and overlap with engineering education themes [37, p. 86]

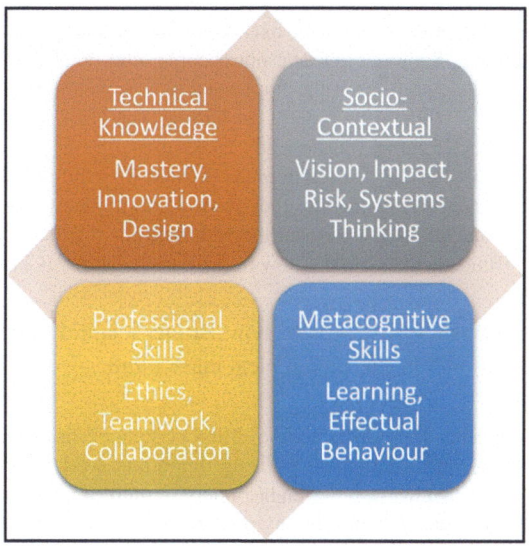

were first considering leadership integration into engineering at UTEP, Dr. Vinod Lohani from the Department of Engineering at Virginia Tech presented a research seminar to our faculty and introduced us to a concept that has become central to our BS in Engineering Innovation and Leadership (E-LEAD): the spiral structure for curriculum.

The concept of spiral curriculum was first introduced by Jerome Bruner [4] in his book *The process of education.* Bruner emphasized the importance of a curricular structure that cycles back on itself at increasing levels of sophistication, enabling learners to build on and reapply what they have already learned to increasingly complex situations. Implementing a spiral curriculum requires a complete overhaul of the traditional curricular structure.

Restructuring curriculum is a difficult and time-consuming process. When pressured to reform curriculum (as we are now), making less taxing modifications to existing curriculum may seem like the easiest or most efficient response. Lohani et al. explain:

> Inevitably, many, if not most, reform movements tend to circle back to a few fundamental ideas that have been difficult to dislodge from the public consciousness. Essentially we "know how it should go" and when pressured to find ways to "do it better" it is easy to conclude that the reason for lack of success is that we have simply not enacted the basic moves well enough. ... It is a more difficult proposition to conclude that we really need to do things in a fundamentally different way. (Lohani et al. [34], p. 3])

Those "basic moves" describe a linear and hierarchical structure that orders information from the most basic and simple to the increasingly complex, with most of the practical application and transfer reserved for senior year projects. Clark et al. offer an insightful analogy:

> This process has been likened to the following hypothetical method of training a baseball team. Suppose you take nine people who don't know the game of baseball and train them individually in all the fundamentals for two years; two months on throwing, three months on catching, five months on hitting, etc. Then, without ever having them practice, or even watch a game, you suddenly ask them to play the game properly as a team. Many would likely quit after the first few months because they didn't like throwing the ball over and over when they didn't know why they were doing it. Those that survived the program would probably play well at the end, but they'd have bruises from those first few games when they knew the fundamentals but not how to put them together. (Clark et al. [8, p. 224])

Lohani et al. explain how Bruner's spiral curriculum design offers an alternative to traditional linear, hierarchical curriculum:

> Bruner's argument took a different stance—proposing that education should involve early legitimate participation in the important work of any discipline of choice. Bruner's idea was that learners—even beginners—could engage successfully with the central problems and questions inherent in any discipline if those key questions could be represented in a manner that invites real experimentation and inquiry at the appropriate level. One key to this idea is that the learning curriculum could be arranged so that the central questions, or themes in a discipline, would be returned to again and again as learners advance in their knowledge and intellectual capacity. The learning trajectory is thus represented as a spiral rather than the linear pathway that is characteristic of traditional schooling. ... As learners participate in increasingly complex investigations, organized carefully around the major themes of choice, they acquire in a more natural way the knowledge they need because it is connected to problems of real import and interest, and they acquire also the full intellectual apparatus associated with *being* [emphasis added] the scientist, historian, or engineer rather than *learning about* [emphasis added] their chosen discipline. In particular, it is this notion of learning *to be something, rather than learning about something* [emphasis added], that we saw as a key basis for reformulating our curriculum. If the curriculum includes early opportunities for prospective engineers to engage in solving problems that are representative of engineering work, albeit at introductory levels, then students would more likely move toward aligning their identities with the profession. (Lohani et al. [34, p. 4])

A spiral curriculum structure recognizes that learners' understanding is often incomplete or even incorrect. It leads learners to double back on what they have learned, developing and correcting as they apply concepts to progressively more complex problems and situations. Spiral curriculum has become a core element of our E-LEAD program at UTEP.

I like to explain our spiral curriculum as "time in the oven." Suppose a baker wants to bake a cake, using a recipe that calls for 30 min in the oven at 350°. That specific combination of time and temperature allows the cake to rise and cook thoroughly. But a novice baker in a hurry could reason that since the cake is the product of time and temperature, they can cut the time down to 1/3rd and triple the temperature (1050 degrees!!). What would the cake look like after 10 min in the oven? The outside would be burned and the inside raw.

Unfortunately—and oftentimes—we take a similar approach to education: cram in all the information and accelerate the time. However, students need to work through projects in bite-sized chunks, learning and re-learning different parts of the process through repetitive iterations. The curriculum spiral allows each level of learning to have "more time in the oven" as students learn, apply, revisit and reapply what they have learned to increasingly complex situations as they progress up the spiral.

Figure 2 illustrates the professional development and leadership topics that are integrated into the engineering curriculum in the E-LEAD program. Each of the eight topics is included every year, with varying amounts of emphasis based on student needs and academic year. (For example, greater attention is given to communication in year one and to innovation in year three, but all elements are addressed to some degree every year.)

Lohani et al. point out that learning is situated in specific contexts:

> Developments in learning research in just the past two decades have moved beyond the primary emphasis on cognition and memory to focus on learning as a culturally, socially situated phenomenon. This research emphasizes that learning always occurs in some type of context

Fig. 2 UTEP spiral curriculum

or culture. Classrooms and campuses represent a distinct type of culture for student learners
.... (Lohani et al. [34, pg. 5])

I want to discuss the E-LEAD program at UTEP as an example of a situated leadership integration using a spiral curriculum. The program was developed in collaboration with Olin College. The goals of our collaboration include our hope to provide "mentorship to other engineering programs interested and ready to adopt new approaches to engineering education" [20, p. 1]. My hope is that this discussion will be helpful to others who are exploring ways to include leadership in engineering education. Because learning is socially and culturally situated, the cultural and socio-economic contexts of UTEP are important to understanding the E-LEAD program.

UTEP is part of The University of Texas system. The university is located in the high desert of West Texas, directly alongside the borders with Mexico and the state of New Mexico. Four factors combine to distinguish UTEP from most other US universities.

First, UTEP is one of less than 2% of US colleges and universities with an open access admissions policy. Open access means that any student who graduates from high school and applies to UTEP is accepted; no entrance test scores, recommendations, or writing samples are required. In fact, El Paso public schools often *require* students to apply to UTEP in their senior year. Of course, students are not required to attend UTEP, but knowing the option is available helps motivate them to continue their education.

Second, because of our location and open access, the profile of our students is unique (see Table 1).

I think it is noteworthy that while serving a bi-racial, low-income community, UTEP is a Carnegie 1 research university with a research portfolio of over $146 million per year and growing. This supports the statement made by our former UTEP President Diana Natalicio: "You can have both access *and* excellence."

The third factor is UTEP's focus on student assets. Most universities operate from a needs-based approach, looking at deficits in students' abilities, knowledge, and skills. Alternatively, UTEP takes an asset-based approach, focusing on the assets students bring to their education (see Table 2).

Table 1 UTEP student profile

80% Hispanic	70% working while attending school (many attending part-time because they must contribute to household income)
85% from El Paso County	95% commute (generally between 10 and 40 miles per day)
5% Mexican nationals	73% Pell Grant eligible (financial need)
51% first generation college students	37% from households with incomes below the federal poverty line

Table 2 Contrast of typical needs-based approach and UTEP's asset-based approach

Needs based	Asset based
Focuses on an imposed standards and deficits from educational background, test scores, etc.	Focuses on students' existing capacity and resources they can apply to enhance their education
Views community members as having something done to them	Values community members as assets/contributors
Reactive	Proactive involvement in community and in students' and community members' lives
Sees community as in need of external experts	Acknowledges community as experts, valuing their knowledge from growing up in the community and learning how to live and survive

Most of our students bring the following assets to their educational journey:

- Bilingual and biliterate in Spanish and English
- Binational orientation—most students regularly cross the US-Mexico border
- Highly engaged in their families and in the community
- Globally aware; students understand what is happening in the US and in Latin America and understand the differences between nations and states
- Resourceful—because they tend to be low income, they have learned how to make things happen and solve problems without having the income to spend money on a solution
- Practical—have learned practicality out of necessity
- Industrious—willing to work hard; have learned to balance work and education
- Entrepreneurial—self-motivated to better their lives and the lives of their families.

This focus on assets instead of needs differentiates UTEP from many institutions across the United States. Engaging these student assets enables us to motivate and involve students in an education that will give them a better future.

Finally, UTEP takes a three-part approach to student success. We have *academic outcomes* that define what we want our students to achieve and *institutional outcomes* that we want to achieve as a university. Then we blend those outcomes into *holistic outcomes* so that students not only acquire the knowledge and skills of an academic education but also gain the ability to enhance their own skill sets and natural assets to benefit their communities after graduation.

E-LEAD Program Development
Since 1996, ABET, the National Academy of Engineering, the National Science Board, and the Carnegie Foundation for the Advancement of Teaching have called for curricular reforms to prepare engineers with the skills necessary to meet contemporary and future

engineering challenges. That call was elaborated in 2004 with the National Academy of Engineering's publication of *The engineer of 2020: Visions of engineering in the new century*. In 2007, UTEP engineering faculty began responding to that call by researching and planning curriculum that would prepare graduates for the current and future workforce.

As planning continued, UTEP hosted an Engineering Lecture Series in which renowned speakers shared their insights. Faculty participated in workshops that gave them opportunities to ask questions and contribute ideas toward the proposal of a new engineering degree offering. A 2011 canvas of student interest demonstrated the new program's viability, and in 2013, the UTEP College of Engineering formed a partnership with Olin College, a highly respected private institution known for its outstanding and innovative program in engineering. One part of this collaboration included UTEP engineering faculty taking residencies at Olin College, learning their pedagogies and approaches from the inside.

The Texas Higher Education Coordinating Board approved the degree plan in 2014, and in 2020, the BS in Engineering Innovation and Leadership—E-LEAD—became the first program of its kind to be fully accredited by ABET. Figure 3 shows the three pillars of leadership in the E-LEAD program.

Over the last eight years, we have graduated over 100 students in the E-LEAD program. A recent alumni survey asked graduates to rate how well their E-LEAD education prepared them for 15 leadership competencies (e.g., initiative, character development, development of inventive solutions, self-directed and team learning, etc.). Responses to all 15 competencies rated between 4.12 and 4.60 out of 5. A second question asked how well their education prepared them for 12 skills and competencies in their current position (e.g., initiative, problem solving, working independently, written communication, etc.). Cumulative scores rated 10 of the 12 preparedness areas between 4.08 and 4.52 out

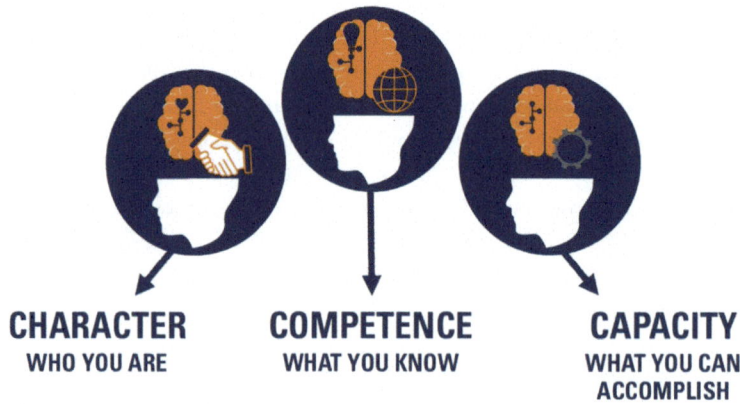

Fig. 3 The three pillars of leadership used in E-LEAD

of 5. (Dealing with cross-cultural and gender differences scored 3.92; training employees scored 3.76.) The following summary findings are based on student comments:

- Alumni expressed gratitude for the E-LEAD program, highlighting its role in preparing them for various challenges and situations.
- Alumni highlighted their ability to apply engineering knowledge in interdisciplinary settings and communicate complex concepts effectively to non-technical stakeholders.
- Alumni shared their experiences of initially feeling disconnected but later finding a sense of belonging and identity within the program. Mentorship from faculty was highly valued, and there were calls for continued support and guidance from knowledgeable and empathic mentors.
- Alumni gained confidence in problem-solving and leadership roles through the program and appreciated the program's emphasis on learning and finding efficient solutions.

We are proud of the E-LEAD program, but as previously noted, all of us who are working in the area of engineering leadership still have much to learn. We are using alumni feedback as we continually seek to improve the E-LEAD program.

Developing the E-LEAD program took eight years. Another seven years passed before the degree received ABET accreditation, and we are constantly researching, collaborating, and learning how to improve E-LEAD. We are proud that all of our graduates are either employed or have advanced to graduate school. Top technology firms seek out our graduates and increasingly recruit them prior to graduation. We are also proud that on average, 20% of UTEP's engineering graduates are women. This percentage is significantly higher than the National Science Foundation's report that just 12% of employed engineers in the US are women [52].

Broadening Access and Participation in Engineering Education

Both the UK and the US are facing a critical shortage of engineers. Research from Stonehaven, a London-based consultancy firm, indicates that as many as 20% of current engineers in the UK are planning to retire within the next few years, and too few engineering students are being recruited and trained. The result is an anticipated "shortfall of one million engineers by 2030 which runs the risk of delaying vital infrastructure projects" [50]. A similar crisis is predicted in the US: the shortfall of engineers is expected to reach 6 million by 2026 [3]. To address these shortfalls, universities in the UK and the US must work to broaden access to our programs by helping students develop identity, supporting diversity, and removing barriers that block students from receiving an engineering education.

Identity

Hughes et al. explain the vital role of identity in learning:

> Identity is central to the learning process—engaging in a practice and becoming a member of a community of practice lead to a transformed sense of self relative to that practice. ... Learning a practice is thus a social process situated within the context of the community of practice .... (Hughes et al. [26])

Experience has taught me that the two most important factors for first-year students are intrinsic motivation and self-identity. At its core, self-identity grows out of a student's familial, social, economic, and cultural situatedness. Self-identity includes the student's sense of self-worth and self-confidence—critical components in student motivation and success. Students must want something and see themselves as being able to attain that something. This is one reason why the assets-based approach at UTEP is so important. As previously noted, our students are largely Hispanic (Mexican descent) and often grow up in poverty. Over half are first-generation college students, so they lack familial or community role models for higher education and professional careers. By focusing on assets, we nurture students' self-identity.

In the E-LEAD program, we help first-year students explore their self-identities by answering some specific questions: Who are you? How do you express who you are? With whom do you identify? As part of this exercise, we give each student an $18'' \times 18''$ piece of platform on which to construct a physical model of their identity. Two student projects are pictured in Fig. 4.

After students share their projects, all E-Lead faculty members share their personal journeys of self-identity and what motivated them to develop an engineering and a leadership identity. Then each student is assigned to create a graphical representation that builds motivation by expressing who they want to become and explores how they can move from who they are now to the person they want to become. These two exercises help students understand and value their own self-identities and understand what motivates them to become successful engineers.

Unfortunately, sometimes engineering elducators work on helping students develop an engineering identity before those students have established a self-identity of worth and value. A strong self-identity is foundational to developing any professional identity.

Developing an engineering identity is an important second step because it facilitates learning and contributes to persistence while studying engineering. Loui and Borrego found that "Students who identify with engineering, or who view their own identities as consistent with engineering, are more likely to select engineering as a career and to persist through engineering programs." They also note that engineering has often been described as a "leaky pipeline," citing research findings that "among those who start as first-year students in engineering, only 57% persist in engineering to the eighth semester" [35, pp. 6, 15]. An engineering identity gives students the resilience to persist and enables them to see themselves as part of a supportive engineering community. Identifying as an

Fig. 4 Student projects exploring identity. The note attached to the project on the right reads as follows: "The sculpture itself, the book, represents my love for reading. The pages are divided by color and order for a specific reason. The black pages represent the past. They are black because we can no longer go to the past, but we can learn from it. The present pages are blue, and I choose to put my storyboard to represent what my present looks like. Lastly, the future pages are white to give them a 'blank' look because we can still control what the future looks like and work towards having a bright future, hence the lights."

engineer and as part of an engineering community motivates students to continue learning and studying when their friends from high school pressure them to hang out or when they are challenged by families who may lack confidence in their ability to succeed or pressure them to work full-time to help support the family. A UTEP study confirms that a student's participation in experiences outside the classroom—engaging in an engineering community—is the most accurate indicator of persistence.

Most of the programs I have reviewed do not emphasize the importance of engineering identity until students are near the end of their academic careers. They begin by emphasizing knowledge, then skills, and finally identity and community. I think we have it backwards. Figure 5 illustrates the contrast between the most common order of emphasis and the importance of these key factors.

Developing an engineering identity is an important component of a student's education, but to be prepared for the demands of industry, it is equally crucial to help students develop a leadership identity. Schell and Hughes [49] explain these two facets of identity: engineering identity, "including academic and professional factors, is shown to improve persistence … and create a greater sense of belonging in the field." Leadership identity includes students' developing "a personal sense of identity as a leader as they deepen

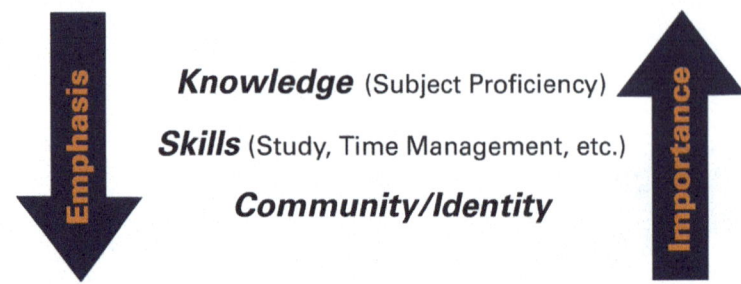

Fig. 5 Emphasis versus importance of key factors in engineering education

their understanding of what constitutes leadership—namely, that leadership is a process, rather than a position" [49, p. 130]. Students who do not self-identify as leaders will rarely act as leaders or take on leadership roles. Lohani notes that the spiral curriculum is a helpful strategy for integrating identity development. From the very beginning, the spiral emphasizes *being an engineer* and *being a leader* rather than just *learning about engineering and leadership* [34, p. 4].

Much of a students' self-identity has already been developed by the time they begin university studies. Middle and high school students study math, physics, biology, chemistry, geography, and computing, but not engineering. They have no formal way to understand what engineering is or what an engineer does. In the UK, the Society of Operations Engineers (SOE) found that young people's perceptions of engineering were often negative: "Two-thirds of people said they had never thought of a career in engineering, even though engineering came out top when people were asked to rank professions they most admired." Note the following research findings:

- Responses from GenZ (born between 1997 and 2012):
  - 30% said they would not choose a career in engineering
  - 34% think engineering is too male-dominated
  - 32% dislike the centrality of math and science in engineering
- Responses from Millennials (born between 1981 and 1996):
  - 19% said they would not choose a career in engineering
  - 28% believe engineering is too male-dominated
  - 28% were unaware of opportunities in engineering [50].

A survey of US youth between the ages of 8 and 17 (Generation Alpha) yielded similar results:

- 44% "don't know much" about engineering
- 30% want a career that is more exciting than engineering

- 21% said they know too little about math and science to be good at engineering (despite the fact that 22% of youth surveyed named math as their favorite subject and 17% ranked science as their favorite) [42].

These studies reveal that in both the UK and the US, too many young people are building self-identities that do *not* include the possibility of engineering. As engineering educators, we need to do a much better job of collaborating with primary and secondary schools to introduce engineering to students in meaningful ways. In *The process of education*, Bruner [4] explains that the benefits of a spiral curriculum are applicable at any age. Even early learners can engage with the central questions of any subject area if those questions are presented in an age-appropriate way and appeal to the child's natural appetite for discovery. But primary or secondary schools have too few teachers who are subject experts in math, science, and physics and inadequate resources to support discovery-motivated learning. If we as engineering educators are to prepare youth to choose engineering and fill the gaps that threaten our infrastructures, we must significantly increase our outreaches into schools and educate teachers who will inspire and motivate our youth to begin forming an engineering identity long before they approach the university.

Diversity

In both the UK and the US, we are extremely unlikely to have an adequate engineering leadership workforce if we do not increase our efforts to recruit, retain, and graduate larger numbers of underrepresented people groups. A report from the Royal Academy of Engineering confirms that "It is vital that the engineering profession draws from the full talent pool to help meet the anticipated skills need." But the RAE offers a second, more practical reason for diversity: "Given engineering's role in shaping the world around us it is vital that the engineering workforce broadly reflects the diversity of society to ensure that what is designed, developed and made meets the needs of the many, not the few" [11]. Yet after decades of efforts to advance gender and racial equality, "on average, globally, women hold 29 percent of executive level leadership roles, while white, middle class men are disproportionately privileged as leaders" [9, p. 233].

In the UK, women comprise 48% of the total workforce but only 16.5% of the engineering and technology workforce [32]. Women also comprise 48% of the total workforce in the US, but the percentage of women in engineering is just 15% [16].

Ethnicity numbers are even more concerning. Ethnic minorities make up 12% of the total labour force in the UK with just 9% of working engineers having a minority heritage [14]. In the US, ethnic minorities make up 39% of the total workforce, and only 13.7% of engineers come from minority groups [12]. Maier, on behalf of the Royal Academy of Engineering, states,

> Firstly, we need to put rocket boosters under our collective efforts to make engineering more inclusive. There is still nowhere near enough young women and girls entering our profession. This also applies to UK Black Asian and minority ethnic (BAME) communities too. We are still not representative of wider society—and that has a knock-on impact when attracting young talent, and why we still do not have enough young people studying STEM subjects. (Maier [36])

Hughes et al. [26] found that gender can complicate identity development for women students since recognition from others—an important factor in becoming part of an engineering community—is more readily granted to men than to women. Coleman's [9] study of women in leadership roles reveals that stepping out of their culturally-socialized roles and into a leadership role requires a shift in traditional identity.

> The main underlying issue for women in leadership is that generally men and women are allocated to different roles in societies, with men being seen as agentic and women more passive and supportive, men operating outside the domestic sphere and women within it. (Coleman [9, p. 235])

Ong et al. report that minorities experience similar struggles. Women of color have reported that they experience "most current engineering educational environments ... as individualistic, isolating, competitive, and chilly because they conform to the culture norms of the White male majority." The authors cite calls to understand the complexity of "interlocking gendered and raced oppressions ... for those who embody multiple marginalized identities" [43, pp. 583, 606].

As educators, we must increase our efforts to support all types of students. Chan et al. emphasize the individual nature of support:

> As engineering educators we have a responsibility to ensure that access to support can be made available to students in a more equitable way. In the same vein, ... [we] need to tailor supports to the recipient, which, interpreted through an equity lens, reminded us that it is not about ensuring everyone has the same supports but facilitating an equitable allocation of supports so all engineering graduates can fulfill their leadership potential. (Chan et al. [6, p. 1264])

Supports often take the form of connection through networking, social interaction, mentoring, sponsorship, and activities. International students and students who lack familial support may need supports that create a sense of belonging. Ong et al. cite a biomedical engineering program that planned biannual outings for students and faculty to interact through sports, game nights, picnics, and even a boat cruise. Students said that these activities "humanized" their professors and created "a whole family feeling" [43, p. 606]. Chan et al. argue that while students need to be made aware of available support, they also need to learn how to "proactively secure support." Successful engineering leaders learn "to figure out their needs in order to locate the supports" that help them achieve their goals [6, p. 1258].

Barriers to Access

Barriers can range from standards a student must meet, to program structures, to lack of financial resources and many more. In the first essay of this series, we looked at key differences in the structure of engineering programs in the UK and the US. These differences affect students' acceptance to university as well as their ability to attend school part time, work while attending school, and change majors. More research is needed to assess how our programs can be structured to allow broader access to an engineering education.

One of the most significant access barriers is entrance test requirements. Access to engineering programs has long been based on the belief that test scores are reliable predictors of success and should, therefore, be used as criteria for admission. UK academics accept students into a department based on A-level scores, and US universities have long relied on ACT or SAT scores.

Both the UK and the US have at least some options available to students whose scores do not qualify them for admission to a university program. In the UK, the main pathway is a foundation year course that helps students gain subject knowledge and prepare for studying at the university level. The number of students taking foundation courses has increased 700% over the last ten years [7], but foundation courses have also come under some harsh criticism. "Less than three-quarters of students (74%) proceed from the foundation year to a full degree course or get a qualification, compared with 91% of undergraduate students ... that makes 19,000 students a year whose foundation year does not even get them into degree-level study ...." [19]. Foundation courses have also been attacked for inconsistent quality, and universities have been charged with using them to attract international students who have lower qualifications than home students but who pay higher fees. Many foundation courses are still fairly selective, and students incur additional expense along with the extra year of time. Additional research is needed on the effectiveness of foundation courses and their efficacy as a strategy to remove access barriers.

Over the last few years, more than 80% of four-year institutions in the US have eliminated entrance test score requirements [40]. However, renowned institutions such as Harvard, Yale, MIT, Georgetown, and Dartmouth along with all public universities in Florida and Georgia have reinstated test score requirements [24]. Many universities across the country are expected to follow. At this point, though, many schools are "test optional," basing admission decisions on other factors that may or may not more accurately predict student success. These include recommendations from teachers and counselors, personal essays, and interviews along with grades and engagement during high school.

UTEP commissioned a Lumina Research Study to analyze what factors *actually do and do not* predict success for our students (with success defined as passing classes and graduating with a degree). The most important factors that do predict success are preparation and engagement: high school GPA and class rank, attendance, and engagement in school activities such as sports or clubs. The study also illuminated factors that do not reliably

B	The leaders in equal access to higher education			
	These are the public research universities with the most low-income students			
No.	University	Median Family Income	Share of students from the top 20%	Share of students from the bottom 20%
1.	University of Texas at El Paso	$42,400	11.4%	28.0%

Fig. 6 Brookings institute report

predict success. Ethnicity, household income, parental education levels, and ACT/SAT test scores are not reliable predictors of success, at least not for our students at UTEP [5].

The Lumina study also identified some risk factors that impact student success. For example, failing a course, especially in the first term, significantly decreases a student's likelihood of success. External demands on a student (e.g., having to work more than 20 hours a week) also have a negative impact [5]. Risk assessment remains a research challenge for engineering educators. We have not yet found an effective tool or method for assessing a student's risk factors, but when we can identify a risk factor, we try to address it.

For example, most of our students have to work. Driving to and from campus to a job takes time out of a student's day and also adds gas and vehicle expenses that complicate financial need. We have worked very hard and invested significant funds to provide undergraduate student employment opportunities on campus so that students can work and go to school without the extra time and expense of travel.

We have a lot more to learn, but we are excited about the successes we have seen. The Brookings Institute named UTEP #1 in the US in equal access to higher education [23] (see Fig. 6).

It would not be unusual to expect that since UTEP is an open access university, our graduation rates would be low, but graduation data tells a different story. Universities that are minimally selective or nonselective have an average graduation rate of 27.1%. The most selective universities with the strictest criteria for admissions have an average graduation rate of 82.8%. UTEP's graduation rate hovers between 64.7 and 67.9%, significantly closer to the rate of the most selective universities than to the minimally selective or nonselective institutions [21]. We believe this level of success is a direct result of UTEP's asset-based approach to education and our commitment to student success (Fig. 7).

What Comes Next?

Engineering education programs in both the UK and the US must work diligently to redefine engineering education to address the growing need for qualified engineers that meet the demands of modern industry. Universities must develop new strategies for incorporating professional development, especially the leadership skills industry demands. The

Fig. 7 Comparison of graduation rates for UTEP (open access) and other universities based on selectivity of admissions

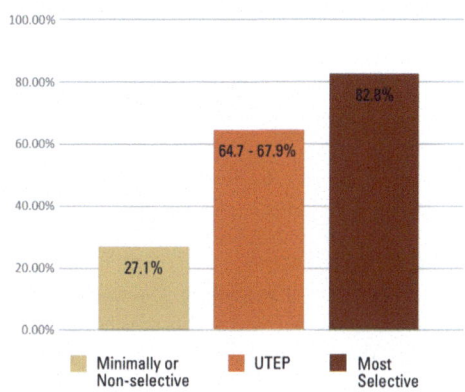

most effective strategy is to integrate leadership development into existing engineering curriculum using a spiral curriculum model that allows students to revisit material as they apply it to progressively more complex situations.

Universities must also broaden access to engineering programs. A critical component of this broadening is to create a culture that emphasizes identity and motivation beginning in the first year. After nurturing a strong self-identity, we must help students develop an engineering identity and a leadership identity. When students develop a sense of belonging in the engineering community, they are more likely to persist in their education, especially when facing academic, social, or financial challenges. Students who can envision themselves as leaders are more likely to develop the leadership skills needed for success in their careers.

Finally, universities must continue to identify and remove barriers to engineering education. One of the most significant barriers is the reliance on standardized test scores as predictors of potential success. Research indicates that these scores are not a reliable predictor of success, especially when compared to factors such as a student's demonstrated engagement and persistence during secondary school. Broader access depends on the development of new measures that more accurately predict which students will thrive in an engineering program.

References

1. ABET. Criteria for accrediting engineering programs, 2022–2023. (2021). *ABET*. https://www.abet.org/accreditation/accreditation-criteria/criteria-for-accrediting-engineering-programs-2022-2023/
2. Blundell, N. (2019, 27 December). A century of man-made disasters. *Pen and Sword History*.

3. Bridging the gap: Addressing the critical shortage of U.S. engineers. (2024, 19 March). *Lighthouse Professional Services*. https://www.lighthouseprofessionalservices.com/blog/bridging-the-gap-addressing-the-critical-shortage-u-s-engineers-178
4. Bruner, J. (1960). *The process of education*. Harvard University Press.
5. Center for Institutional Research and Planning. (2007, 18 May). Access vs. success: Preliminary policy insights from a study funded by the Lumina Foundation for Education. *The University of Texas at El Paso*. https://www.utep.edu/planning/cierp/_Files/docs/vpba-project/Access-v-Success.ppt
6. Chan, A., Rottmann, C., Reeve, D., Moore, E., Maljkovic, M., & Radebe, D. (2023). Making the path to engineering leadership more equitable: Illuminating the (gendered) supports to leadership. *European Journal of Engineering Education*, 48:6, 1249–1268, https://doi.org/10.1080/03043797.2023.2272819
7. Chowen, S. (2023, 19 October). Uni foundation year numbers up 700% in 10 years. *FEWeek*. https://feweek.co.uk/uni-foundation-year-numbers-up-700-in-10-years/
8. Clark, W. M., DiBasio, D., & Dixon, A. G. (2000). A project-based, spiral curriculum for introductory courses in CHE: Part 1. Curriculum design. *Chemical Engineering Education*, 34(3), pp. 222–228. https://journals.flvc.org/cee/article/view/123056
9. Coleman, M. (2019). Women leaders in the workplace: Perceptions of career barriers, facilitators and change. *Irish Educational Studies*, 39(2), 233–253. https://www.tandfonline.com/doi/full/https://doi.org/10.1080/03323315.2019.1697952
10. Daley, J., & Baruah, B. (2020). Leadership skills development among engineering students in higher education—An analysis of the Russell Group universities in the UK. *European Journal of Engineering Education*. 10.1080.03043797.2020.1832049
11. Education for Engineering. (2019). Engineering skills for the future: The 2013 Perkins review revisited. *Royal Academy of Engineering*. https://www.raengorg.uk/perkins2019
12. Engineer demographics and statistics in the US. (2024). *Zippia*. https://www.zippia.com/engineer-jobs/demographics/
13. Engineering Ethics Reference Group. (2022, February). Engineering ethics: Maintaining society's trust in the engineering profession. *Royal Academy of Engineering, & Engineering Council*. https://www.engc.org.uk/media/3921/engineering-ethics-report-february-2022.pdf
14. EngineeringUK. (n.d.). EngineeringUK briefing: Social mobility in engineering. *Engineering UK*. https://www.engineeringuk.com/media/1762/social-mobility-in-engineering.pdf
15. Epstein, D. (2024, 03 March). Quality vs. quantity in engineering. *Inside Higher Ed*. https://www.soc.duke.edu/GlobalEngineering/pdfs/media/FramingEngineering/InsiderHigherEd_QualvsQuan.pdf
16. Evans, C. (2024, 02 January). Women in engineering statistics: 32 notable facts. *Fictiv*. https://www.fictiv.com/articles/women-in-engineering-statistics-32-notable-facts#:~:text=By%202019%2C%20women%20represented%2048,1970%20to%2027%25%20in%202019.
17. Fleet, D. (2016, 15 April). The last messages from *Titanic*. *The Citizen Online* (D. Fleet, Ed.). https://thecitizenonline.com/the-last-messages-from-titanic/
18. Fredde, A. (2021, March 10). Engineer who refused to approve *Challenger* launch dies in Ogden. *KSL*. https://kslnewsradio.com/1944652/challenger-engineer-dies/
19. Freeman, J. (2024, 07 March). The proliferation of foundation year courses has created some blind spots. *Wonkhe*. https://wonkhe.com/blogs/the-proliferation-of-foundation-year-courses-has-created-some-blind-spots/
20. Golding, P., Gonzalez, R., Moreno, G., Schoephoerster, R., Starks, S., Vaughan, M., & Townsend, J. (2015). *The creation and inauguration of engineering leadership: UTEP and Olin College innovation project*. (Paper presentation). 2015 IEEE Frontiers in Education Conference.

References

21. Gonzalez, R. V. (2023). *Who gets to graduate? Access to HE: UTEP and regional student success.* CEE conversation series. [Video]. YouTube. https://www.youtube.com/watch?v=RZeg3kqUPcw&t=35s
22. Grimson, J. (2002). Re-engineering the curriculum for the 21st century. *European Journal of Engineering Education,* 27.1. https://doi.org/10.1080/03043790110100803
23. Halikias, D., and Reeves, R.V. (2017, 11 July). Ladders, labs, or laggards? Which public universities contribute most. *Brookings.* https://www.brookings.edu/articles/ladders-labs-or-laggards-which-public-universities-contribute-most/
24. Horizon Education. (2024, 03 May). List of colleges dropping & reinstating ACT/SAT requirements. *Horizon Education.* https://horizoneducation.com/blog/colleges-dropping-reinstating-act-sat-requirements
25. Howell, E., & Dobrijevic, D. (2023, 25 January). *Columbia* disaster: What happened and what NASA learned. *SPACE.com.* https://www.space.com/19436-columbia-disaster.html
26. Hughes, B. E., Schell, W. J., & Tallman, B. (2018). Understanding engineering identity in undergraduate students. In *Proceedings of the American Society for Engineering Management 2018 International Annual Conference.* Ng, E-H., Nepal, B. Schott, E., & Keathley, H. (Eds.) https://par.nsf.gov/servlets/purl/10089982
27. Kendall, M. R., Chachra, D., Gipson, K., & Roach, K. (2022). Motivating the need for an engineering-specific approach to student leadership development. In M.R. Kendall & C. Rothmann (Eds.). *New Directions for Student Leadership: No. 173. Student leadership in engineering* (pp. 13–21). Wiley. https://doi.org/10.1002/yd.20475
28. Khalid, A., Chin, C.A., Atiqullah, M. M., Sweigart, J. F., Stuzmann, B., & Zhou, W. (2013). Building a better engineer: The importance of humanities in engineering education. *ASEE PEER.* Article 6052. https://peer.asee.org/building-a-better-engineer-the-importance-of-humanities-in-engineering-curriculum.pdf
29. Klassen, M., Reeve, D., Rottman, C., Sacks, R., Simpson, A. E., & Huynh, A. (2016, 26–29 July). *Charting the landscape of engineering leadership education in North American universities.* (Paper presentation). ASEE 123rd Annual Conference & Exposition. https://tspace.library.utoronto.ca/bitstream/1807/77374/1/charting_landscape_engineering_leadership_programs.pdf
30. Koromyslova, E., Steinlicht, C., & Peter, M. K. (2023). Closing the professional skills gap for engineering graduates: Recent trends in higher education. *ASEE.* https://peer.asee.org/closing-the-professional-skills-gap-for-engineering-graduates-recent-trends-in-higher-education.pdf
31. Kovarik, B. (n.d.). Radio and the *Titanic. Revolutions in ommunication: Media history from Gutenberg to the digital age.* https://revolutionsincommunication.com/features/radio-and-the-titanic/#:~:text=Technically%2C%20the%20problem%20with%20the,other%20ships%20within%20signaling%20distance
32. Leevers, K. (2023, 08 March). IWD—Why we need more women in engineering and technology. *EngineeringUK.* https://www.engineeringuk.com/blog/iwd-why-we-need-more-women-in-engineering-and-technology/
33. *Letters of Note.* (2010, 28 June). https://lettersofnote.com/2010/06/28/we-are-sinking-fast/
34. Lohani, V K., Wolfe, M. W., Wildman, T., Mallikarjunan, K., & Conner, J. (2011, Summer). Reformulating General Engineering and Biological Systems Engineering Programs at Virginia Tech. *Advances in Engineering Education.* ASEE. https://advances.asee.org/wp-content/uploads/vol02/issue04/papers/aee-vol02-issue04-p11.pdf
35. Loui, M. C., & Borrego, M. (2019, 01 April). Engineering education research. In S.A. Fincher & A. V. Robins (Eds.), *The Cambridge handbook of computing education research.* http://www.cambridge.org/9781108721899

36. Maier, J. (2019, January). Engineering 4.0—What next for engineering skills? (Foreword). In Engineering skills for the future: The 2013 Perkins report revisited. Royal Academy of Engineering. https://raeng.org.uk/media/hn4hdep3/perkins_report_jan19_final-web.pdf
37. McDonald, J. R., & Jamieson, M. V. (2022). Diversity of engineering leadership program design. In M.R. Kendall & C. Rothmann (Eds.). *New Directions for Student Leadership: No. 173. Student leadership in engineering* (pp. 83–91). Wiley. https://onlinelibrary.wiley.com/toc/23733357/2022/2022/173
38. Marguiles, C. (2012, 23 October). I didn't see the iceberg!: And other *Titanic* communication mistakes. [Paper presentation]. *PM® Global Congress 2012*—North America, Vancouver, British Columbia, Canada, Newtown Square, PA: Project Management Institute. https://www.pmi.org/learning/library/didnt-see-iceberg-titanic-communication-mistakes-6026
39. National Academy of Engineering. (2004). The engineer of 2020: Visions of engineering in the new century. *National Academy of Sciences*. National Academies Press. https://doi.org/10.17226/10999 https://nap.nationalacademies.org/download/10999
40. Neitzel, M. T. (2023, 13 June). The test-optional college admissions movement continues to grow. *Forbes*. https://www.forbes.com/sites/michaeltnietzel/2023/06/13/the-test-optional-college-admissions-movement-continues-to-grow/?sh=14a09ed81326
41. Newman, M. E. (2019, 13 February). NIST reveals how tiny rivets doomed a titanic vessel. *NIST Time Capsule*. https://www.nist.gov/nist-time-capsule/nist-beneath-waves/nist-reveals-how-tiny-rivets-doomed-titanic-vessel
42. Noria Corporation. (n.d.). Survey shows kids saying no to engineering career. *Machinery Lubrication*. https://www.machinerylubrication.com/Read/1677/survey-shows-kids-saying-no-to-engineering-career
43. Ong, M., Jaumot-Pascual, N., & Ko, L. T. (2020, 10 January). Research literature on women of color in undergraduate engineering education: A systematic thematic synthesis. *Journal of Engineering Education, 109*. https://onlinelibrary.wiley.com/doi/epdf/https://doi.org/10.1002/jee.20345
44. Polmear, M., Volpe, E., Simmons, D. R., Clegorne, N., & Weisenfeld, D. (2022). Leveraging faculty knowledge, experience, and training for leadership education in engineering undergraduate curricula. *European Journal of Engineering Education. 47*(6), 950–969, https://doi.org/10.1080/03043797.2022.2043243
45. Riemer, M. (2003). Integrating emotional intelligence into engineering education. *World Transactions on Engineering and Technology Education*, 2(2), pp. 189–194. https://www.researchgate.net/publication/299566618_Integrating_emotional_intelligence_into_engineering_educationF0aW9uIn19
46. Rottmann, C., & Kendall, M. R. (2022). Looking to the future: Four key purposes of engineering leadership education. In M. R. Kendall & C. Rottmann (Eds.). *New Directions for Student Leadership: No. 173. Student leadership development in engineering* (pp. 149–155). Wiley. https://doi.org/10.1002/yd.20486
47. Rottman, C., Reeve, D.W., Sacks, R., & Klassen, M. (2016). An intersubjective analysis of engineering leadership across organizational locations: Implications for higher education. *Canadian Journal of Higher Education*. https://journals.sfu.ca/cjhe/index.php/cjhe/article/view/186198/pdf
48. Rottman, C., Sacks, R., & Reeve, D. (2015). Engineering leadership: Grounding leadership theory in engineers' professional identities. *Leadership*. https://tspace.library.utoronto.ca/bitstream/1807/77544/3/engineering_leadership_orientations.pdf

References

49. Schell, W. J., & Hughes, B. E. (2022). Developing an engineering leadership identity. In M.R. Kendall & C. Rothmann (Eds.). *New Directions for Student Leadership: No. 173. Student leadership in engineering* (pp. 129–137). Wiley. https://onlinelibrary.wiley.com/doi/abs/https://doi.org/10.1002/yd.20484#:~:text=Their%20work%20found%20three%20orientations,(3)%20organizational%20innovation%20%E2%80%93%20entrepreneurial
50. Society of Operations Engineers. (2023, 03 October). UK faces shortfall of 1 million engineers by 2030. *SOE.* https://www.soe.org.uk/resources/uk-faces-shortfall-of-1-million-engineers-by-2030.html#:~:text=The%20UK%20faces%20a%20shortfall,crucial%20infrastructure%20such%20as%20hospitals.
51. Tikkanen, A. (2024, 28 March). *Titanic. Time Magazine/Britannica.* https://www.britannica.com/topic/Titanic
52. Welker, B. (2024, 17 July). 100+ Important Engineering Statistics for 2024. *Crush the PM Exam.* https://crushthepmexam.com/important-engineering-statistics/#:~:text=As%20of%202023%2C%20the%20total,women%20(National%20Science%20Foundation)

Engineering Education in the Age of Accelerations

So far, the content of this book drew heavily on my observations from a year-long research fellowship, during which I conducted over 200 interviews with academics from nearly 30 institutions in the UK as well as my experience in the engineering industry, as an entrepreneur, as a long-time engineering educator in the US, and as an engineering program evaluator for the Accrediting Board for Engineering and Technology (ABET). How engineering education may be impacted by rapid global changes is explored next. My purpose is to ask how we can, should, and perhaps even must make changes to engineering education not only in the next few years but also, more critically, in the next few decades and beyond.

In 1965, Gordon Moore, who co-founded Intel in 1968, predicted that the capability and speed of computers would double every year as greater numbers of transistors could be contained in a microchip. "Moore's Law" became a "roadmap that the semiconductor industry imposed on itself, driving future development" [21, p. 30].

Initially, Moore predicted that by 1975, a single silicon chip, one inch square (6.25 cm^2) would be able to contain 250,000 components. By 2019, Intel was packing over 100 million transistors onto a chip only one square millimeter in size. Pat Gelsinger, Intel's current CEO, predicts that by 2030, Intel will produce a chip with one trillion transistors [11].

Moore revised his prediction in 1975 to forecast that "chip densities would double every two years, rather than every year" [21, p. 30]. This apparent slowing has led many to proclaim that Moore's Law is no longer relevant, but Gelsinger has stated that Moore's Law is "alive and well." Friedman [11] summarizes Gelsinger's comments:

Gelsinger said during his talk, "I think we've been declaring the death of Moore's Law for about three to four decades." And while that might be true, he did admit that "we're no longer in the golden era of Moore's Law, it's much, much harder now, so we're probably doubling effectively closer to every three years now, so we've definitely seen a slowing." Intel's CEO is pushing a "Super Moore's Law" concept based on using 2.5D and 3D chip packaging to increase transistor counts. Gelsinger also refers to this as "Moore's Law 2.0." (Friedman [11])

Thomas [12], author of *Thank you for being late*, points to Moore's Law as one of three types of global accelerations impacting our world:

- Acceleration of technology
- Acceleration of globalization
- Acceleration of environmental changes

Friedman asserts that "these simultaneous accelerations in the Market, Mother Nature, and Moore's Law together constitute the 'age of accelerations,' in which we now find ourselves" and that they are "transforming almost every aspect of modern life" [12, p. 27–29].

Acceleration of Technology

The history of computers is a story of accelerations. Charles Babbage designed the first computer in 1822, but 123 years passed before the first general purpose, digital electronic computer appeared in 1945. The first personal computer appeared just 28 years later in 1973, and a mere eight years passed before the first laptop became available in 1981. Today, an iPhone 15 has over 14.6 billion times the processing power of the 1969 Apollo 11 Guidance Computer [2, 17]. The cloud has reduced the cost of large amounts of data storage for both businesses and individuals (e.g., Google Drive, iCloud, and Dropbox), and cloud computing processes data much faster than CPUs, again at an affordable rate (web applications, mobile apps, websites) [7].

But those advances barely begin to tell the story. Artificial intelligence (AI) was first conceived mid-twentieth century [29]. In less than 70 years, AI has become ubiquitous. Chat Generative Pre-Trained Transformer (ChatGPT) was released 30 November, 2022. By the end of its first week, it had over a million subscribers [15, p. 2], and subscriptions increased to more than 100 million within two months [18, p. 55]. As of April 2024, Chat GPT had over 180.5 million users worldwide [23].

Chat GPT and other AI applications are already being used in multiple environments for a myriad of applications, including but not limited to the following:

- Education
- Conversation (including language translation and customer service)

- Marketing (content planning and generation [text, image, video], customer assistance, SEO)
- Creative content generation (text, poetry, email, art, image, video, music)
- Coding
- Collaboration and workflows
- Presentations (including slide creation)
- Plagiarism detection (including recognizing content developed through AI) [26].

The estimated value of the global AI market is more than $196 billion, and AI development and use is expected to increase by over 1300% by 2026 [16]. AI puts Moore's Law on steroids.

Acceleration of Globalization

Globalization is changing nearly all aspects of society and life, including markets and economies, wealth, human relationships and communication, work, and education.

> As the world grows more interconnected through the flows of people, goods, and information, many challenges are becoming more difficult to address since human needs are increasingly being met through global supply chains. Global shocks (e.g., war, economic recession, pandemic) can severely disrupt these interconnections and generate cascading consequences across local to global scales. (Viña and Liu [33, p. 95])

Globalization affects wealth in both positive and negative ways. The *2024 Global Wealth Report* predicts that the number of millionaires will increase "in 52 out of 56 developed and developing economies":

> Gains will be led by tech powerhouse Taiwan, where the number of millionaires is set to jump 47% on the back of the booming microchip industry and a rise in immigration by wealthy foreigners. That growth was followed by Turkey (43%), Kazakhstan (37%), Indonesia (32%), and Japan (28%). The two hubs in which the most global millionaires are based, the U.S. and mainland China, are set to see their figures rise 16% and 8% respectively. However, the number of millionaires is forecast to plunge by 17% in the U.K. (Reid [27])

The report also predicts greater inequality of wealth: "A rise in average wealth overlooks a sharp fall in median wealth—implying higher inequality, with wealth becoming more concentrated among the richest" [27].

The COVID pandemic helped to fuel inequalities. Work from home increased from 10.5% to 37.9% for workers with income in the top 10%; for workers in the lower half of income, the rise in work from home was much smaller, increasing from 4.9% to only 11.2% [13]. Post-pandemic, many workers continue to desire and even insist on the greater flexibility of a work from home model, especially since work opportunities have globalized [28].

The same is true of education. Primary and secondary schools along with universities were forced to adapt to social distancing and stay-at-home mandates. That adaptation has led to new and broader use of technology to deliver education over distances. Jaakkola et al. [19] find that education plays a leading role in globalization: education "can be used to promote globalization" and "is also a prerequisite for successful globalization" [19]. Filho et al. note that "HEIs also represent a crucial stakeholder in the promotion and implementation of the United Nations (UN) 2030 Agenda for sustainable development ... and the digitalization of society by producing knowledge for new technologies and social innovation" [10, p. 109].

Acceleration of Environmental Changes

Climate change, population growth, and biodiversity loss form an interconnected set of accelerating challenges. In the US, more than in the UK, climate change has been disputed, but the facts are clear:

> January 2024 was the warmest January on record globally, with an average ERA5 surface air temperature of 13.14°C, 0.70°C above the 1991-2020 average for January and 0.12°C above the temperature of the previous warmest January in 2020. This is the eighth month in a row that is the warmest on record for the respective month of the year. (Copernicus [8])

Both climate change and human migration are accelerating. Adger et al. report that in this century, nearly 800 million people—about 10% of the global population—have migrated within-country from rural to urban settings, adding to urbanization. About one in 30 of the global population have migrated internationally to areas with "large open trading economies" [1, p. 2]. The UN report *World population prospects 2024* describes the reciprocal impact of population growth and climate change:

> While the slow growth or decline of populations is occurring mainly in high-income countries, rapid population growth will occur in low-income and lower-middle-income countries. Specifically, Angola, the Central African Republic, the Democratic Republic of Congo, Niger, and Somalia, very rapid growth is projected, with their total population **doubling between 2024 and 2054** [emphasis in original]. This population growth will increase demand for resources, especially in sub-Saharan Africa and South Asia and, combined with poorly managed urbanisation and rising living standards, it will worsen environmental impacts. Climate change, a major challenge, affects these countries the most, where many rely on agriculture— and food insecurity is prevalent. (United Nations [32])

People are not the only ones at risk; the threat to biodiversity is also accelerating. The UN warns estimates that about one million species currently face extinction [31]. UBS estimates that about 60% of GDP depends at least partially on our ecosystem: "Globally, between USD 235–577bn of crop output is directly attributable to animal pollination" [24].

Friedman [12] quotes Craig Mundie, "supercomputer designer and former chief of strategy and research at Microsoft":

> In the world we are in now, acceleration seems to be increasing. [That means] you don't just move to a higher speed of change. The rate of change also gets faster … And when the rate of change eventually exceeds the ability to adapt you get "dislocation." … "Dislocation" is when the whole environment is being altered so quickly that everyone starts to feel they can't keep up.
>
> This is what is happening now. "The world is not just rapidly changing," adds Dov Seidman, "it is being dramatically reshaped—it is starting to operate differently" in many realms all at once. "And this reshaping is happening faster than we have yet been able to reshape ourselves, our leadership, our institutions, our societies, and our ethical choices."
>
> Indeed, there is a mismatch between the change in the pace of change and our ability to develop the learning systems, training systems, management systems, social safety nets, and government regulations that would enable citizens to get the most out of these accelerations and cushion their worst impacts. (Friedman [12, p. 29])

Engineering Education at the Forefront of Change

While our world is undergoing an acceleration of accelerating change, education tends to change slowly—actually, very slowly. At times, change happens so slowly that education becomes stagnant. Few new teaching approaches, learning strategies, or content changes have been introduced. We tend to use technology primarily to present content and not focus on the technology as itself part of that content.

> In an era of rapid change, standing still is the most dangerous course of action.
> —Brian Kelly

Engineers—and thus, engineering education—must be at the forefront of change. Instead of trying to catch up or even just stay current with global change, engineers must get out in front and operate within the technical and non-technical changes taking place, leveraging them to our advantage. We must *be* the changemakers. That means engineering education programs must help students cultivate leadership and professional skills in order to understand the opportunities and the constraints of the environments in which we work.

These include societal and cultural constraints. Projects can succeed or fail depending on how well we understand the community, culture and people impacted by our work. I have experienced this firsthand in my founding of LIMBS International. Approximately 30–40 million amputees live in developing nations. Only 5% of them have access to prosthetic devices, without which they cannot work or participate in their communities. LIMBS International has provided over 10,000 prosthetics to low-income amputees in more than 50 countries. But we were surprised to learn that some amputees were reluctant to use their prosthetics because their community was suspicious of such devices and shunned people who used them.

Some cultures perceive a physical disability as a curse, the result of an individual's actions that deserve some sort of punishment. In these communities (e.g., in Southeast Asia and parts of Africa), many cultural constraints must be considered to help a handicapped individual embrace technology.

The introduction of technology can often backfire or at least create unintended consequences if the engineer developing and deploying the technology does not understand cultural perceptions and beliefs. Societal and cultural constraints require engineers to consider what modifications to the technology are needed and the extent to which communities and their cultural leaders need education. Every engineer and inventor must grapple with these types of socio-cultural questions and constraints.

Engineers must also work within economic and political constraints. We must understand what factors drive the economic engines of the technological progress we want to make. Those factors vary in different countries and geographies. Political environments can restrain or liberate the goals we want to achieve and the methods we use to achieve them. Political constraints are vastly different even between the UK and the US. One example is the use of technology to reduce climate change. The political structures in the UK are much more amenable to climate change implementations while in the US, the political environment surrounding climate change is highly polarized.

Legal constraints include the role of liability and unforeseen consequences. Liability is an important issue in the use of AI. For example, Grassini highlights the "emergent need to address data privacy and security concerns" in educational use of AI: "Student data's sensitivity and personal nature elevate the risk of data breaches, unauthorized access, and potential misuse of data for noneducational purposes [15, p. 7]. AI pulls content from all available online sources. Is there liability if AI gives copyrighted material to another person without attribution? Who is liable? AI raises challenging questions about ownership of information and knowledge.

Engineers must also work within environmental constraints. What if a new technology adds more carbon to the environment? Is that technology viable? What—if any—tradeoffs do we make between environmental contamination and other aspects of human, animal, or environmental welfare? What criteria should be used to decide such questions?

Engineering is no longer just a technical profession; it is a human and societal profession. It is a leadership profession. Many universities have already expanded their programs to include more professional development and leadership training. But will these additions be enough to prepare engineers for the world in which they will work—not just a few years from now but decades from now?

Artificial Intelligence in Engineering Education

The university must adapt to remain relevant in the future. As Grassini argues,

> Artificial intelligence, with its transformative potential, will substantially influence modern education ... With giants like Microsoft planning to incorporate [CoPilot, its proprietary integration of] ChatGPT across their product range, it is only a matter of time before AI tools become a commonplace fixture in our lives. (Grassini [15, p. 8])

Scholars are actively debating the pros and cons of AI in education. Promising applications of AI for education include its ability to individualize learning. AI can assess a student's learning style, pace, and interests along with subject-matter strengths and weaknesses. Based on this information, it can make real-time adjustments to curriculum and presentation to increase student engagement, remediate learning gaps, and facilitate increased learning. AI capabilities also include immersive learning with augmented reality (AR) and virtual reality (VR). These technologies can make learning more visual and experiential. Another valuable AI strategy is the gamification of education. Engineers tend to be gamers, and AI can generate captivating and entertaining games that deeply engage students in learning.

We must broaden access to engineering education, and AI can help us do that. AI technology can break down geographical, socioeconomic, and language barriers to democratize access to education and enable collaborative learning on a global scale. Students from countries across the globe will be able to learn and work together in an equitable and inclusive learning environment with AI real-time language translation. Grassini notes that AI will also benefit educators, decreasing workloads by generating learning materials and by creating and performing assessments [15, pp. 4].

AI can take over repetitive tasks, but it cannot replace human communication, critical thinking, empathy, or ethics. In fact, AI makes professional skills even more important. Marinkovic [20] cites survey data showing that these and other leadership and professional skills are becoming even more important because of AI: "Far from being peripheral, these human skills are becoming the linchpin in an increasingly interconnected and complex digital landscape."

AI reliability is increasing, but it is far from perfect. Critics have expressed concerns over the fact that it sometimes produces "incorrect or even fabricated information." Educators are also concerned about the degree to which AI will enable plagiarism. Researchers have found that students who use ChatGPT are more likely to plagiarize. Plagiarism detectors have been tested on AI-generated text, and researchers have found that "AI models like ChatGPT can successfully circumvent plagiarism detectors" [15, p. 5].

Researchers have also tested ChatGPT's ability to pass important exams like the Fundamentals of Engineering (FE) exam. They conclude that it is a useful tool for exam preparation, but they are concerned about maintaining the integrity of exam results. Two question sets were used to test ChatGPT's ability to pass the FE exam:

> The first test set comprised verbatim questions, with no modifications except for addressing formatting issues that arose when transferring from the source PDF file. ... The second test set involved non-invasively refined questions, wherein additive guidance was incorporated without altering the question's content. ... the results showed that ChatGPT (GPT-4 Base Model - No Vision) achieved an overall accuracy of 66.42% on the dataset ... When applying a refined approach to the same GPT-4 model, the accuracy improved to 75.37%, which can be reasonably considered a passing grade. (Pursnani et al. [25, p. 5])

Because of concerns like these, the New York City Public Schools and the Los Angeles Unified School District along with two international universities (Sciences Po, Paris, France; and RV University, Bengalaru, India) have banned the use of AI [6]. Others have created strict guidelines for the use of AI, and some are requiring that students take assessments in-person [4]. Regardless of how educators feel about AI, it has already claimed its place in education: developing curriculum, writing and scoring assessments, and generating text and research. AI is here to stay, and we cannot afford to ignore it. We need to exploit its benefits while remaining vigilant about its deficiencies. As Grassini notes, "educational institutions must face considerable challenges in retrospectively implementing policies that foster the safe and effective use of AI tools like ChatGPT" [15, p. 8].

That "safe and effective use of AI" is an important skill that students need to learn. After extensive research, Grassini concludes that "severe measures, such as a complete ban on AI tools like ChatGPT in school and university environments ... may disadvantage students attending schools where these tools are forbidden compared to others" who have access and opportunity to learn the effective use of AI. She also makes an insightful comparison: "banning ChatGPT use for students should be considered equal to banning calculators in math class or banning Google" [15, p. 8].

As industry increasingly capitalizes on AI capabilities,

> preparing students with the requisite skills to thrive in an AI-dominated future is essential. ... By offering students a hands-on experience with these tools, we can foster their understanding and application meaningfully while outlining their limitations and keeping pace with technological advances. ...
>
> It is plausible that in the future, students without training in AI tools could find themselves at a competitive disadvantage in the job market compared to their peers with extensive exposure and practical experience with these tools. ...

Beyond the confines of the classroom, it is crucial to confront and address the potential impact of AI on the digital divide. AI tools could either narrow this chasm by facilitating universal access to learning resources or intensify the divide by disproportionately benefiting those with superior access to technology. We need a cooperative, cross-disciplinary approach to navigate these potential challenges and capitalize on AI's opportunities. (Grassini [15, pp. 8–9])

AI is already changing the landscape of higher education. Universities must evaluate and determine how to embrace that change.

Remote and Individualized Learning
I believe there will continue to be a need and a relevant role for higher education in the future, but it may look very different than what we have now. Students will continue to gather in classrooms, but more of those classrooms may be virtual. Students may attend classrooms and labs from destinations around the globe. AI may simultaneously translate lectures and discussions into multiple languages. To meet global demands and student needs, engineering programs will likely need to be much more individualized. Two students may earn the same degree, but their intellectual and educational outcomes may be achieved through different projects and modules to fit the unique needs of individual students and the geographies and communities in which they will serve. Content may be adapted to aid the learning processes of individual students, remediating learning gaps and challenging students to greater knowledge and broader applications of that knowledge. Universities may need to become more accepting of module credits from other reputable universities. Nearly everything a student needs to learn at the undergraduate level is available online [34]. Some students will bring to the university learning achieved on their own with the help of AI. The university will need to decide how to assess that learning.

Four Functions Higher Education Must Continue to Fulfill

Higher education will need to adapt for any or all of these scenarios, but it will not become obsolete. University engineering education will still be relevant and necessary to perform at least the following functions:

Validation/Certification

First, universities must continue to define and validate engineering competency. Society must be able to trust engineers to solve critical problems that impact safety and quality of life. Engineering errors can be costly and even deadly. Suppose someone who is not university trained researches online and learns the facts for building a bridge. Such a person may claim to be an engineer, may speak the language of engineering, and may be skilled in using AI to help design the bridge. But would any of us take the risk of driving our families across that bridge?

Sometimes, AI is wrong. A generalist might be fooled, but a specialist would not. Universities will still be needed to train specialists. Students need to collaborate, listen, test, and learn from both success and failure in low-risk settings so they learn to assess risk and mitigate failure. They need to learn both the benefits and limitations of using AI. Universities provide the environment in which engineering students gain the practical, hands-on experience that qualifies them to be recognized as specialists.

Universities affirm that engineers are knowledgeable and qualified to do their jobs, that they are worthy of the public's trust. Industry relies on universities to certify that engineers are adequately trained. As AI capability increases to perform more tasks and calculations, human judgment, critical thinking, and decision making will become even more important. The university must continue to teach these skills and to validate or certify that a person is competent to perform the essential and challenging work of engineering.

Structure and Process of Education

Second, universities must continue to guide the structure and process of education. As Bruner wrote in *The process of education*,

> The teaching and learning of structure, rather than simply the mastery of facts and techniques, is at the center of the classic problem of transfer. There are many things that go into learning of this kind, not the least of which are supporting habits and skills that make possible the active use of the materials one has come to understand. If earlier learning is to render later learning easier, it must do so by providing a general picture in terms of which the relations between things encountered earlier and later are made as clear as possible. [5, p. 12]

The structure and process of education are the means through which students learn the knowledge, habits, and skills of engineering and the ability to transfer what they have learned to new and increasingly complex situations. Even in a greatly changed future, universities will still be needed to ensure the structure and process of engineering education.

Mentorship

Third, universities must continue to provide mentorship to engineers. Working engineers cite mentorship as one of the most important elements in their career development: "Mentorship is a cornerstone of professional development in many fields, but it plays a particularly critical role in engineering. This dynamic discipline requires continual learning and adaptation to new technologies and methodologies" [3]. Engineers cannot develop the right solutions if they have not learned to ask the right questions. They need to be guided by someone who imparts not only knowledge but more importantly, wisdom, helping them learn the critical thinking, questioning, and decision-making skills necessary to solve complex engineering problems.

Classroom, lab, and team interaction is essential, but it is not enough. As the role of AI continues to impact our work, we need to nurture the wisdom that grows through personal interaction. Our best model is the ancient Socratic Method of

> shared dialogue between teacher and students. The teacher leads by posing thought-provoking questions. Students actively engage by asking questions of their own. The Socratic Method ... is better used to demonstrate complexity, difficulty, and uncertainty than to elicit facts about the world. The aim of the questioning is to probe the underlying beliefs upon which each participant's statements, arguments and assumptions are built. (The Institute for Teaching and Learning [30])

This kind of mentoring helps mentees understand *why* they think the way they do and empowers them to question themselves—important skills for the risk analysis inherent in engineering work. Both mentors and mentees benefit from the mentoring relationship. That mutual benefit is an important component of the fourth function: community.

Community

Finally, universities must provide community. As discussed in Essay 2 of this series, community is essential to a student's engineering identity and persistence. In the past, communities consisted entirely of physical presence, but technology has changed the ways we experience community. Those of us who are old enough to remember communities of presence may have difficulty understanding and relating to virtual community. Certainly, the virtual cannot and should not completely replace the reality of physical presence, but it can enable the sharing of ideas, concerns, problems, and solutions and facilitate the essential recognition of an engineer's knowledge and skills.

Goodall points out that "Communities share a sense of place. Whether that's a geographical area or a virtual or digital platform." Virtual communities have been developing and expanding since the early 1970s [14]. The accelerating changes in technology (especially AI), markets and climate are profoundly impacting—and in many ways, broadening—contemporary understandings of community. This is an area in which the reciprocal benefit of mentoring is important: we need to learn from our students, who are likely more adept at virtual community than we are. As our classrooms and labs become increasingly global and virtual, we need to invest in developing communities of learning and practice for and with our students.

Final Thoughts

This part of the book looks toward the future of engineering education in light of the rapid and accelerating changes happening in our world. My hope is that in discussing these topics, we will ourselves engage in the Socratic Method of dialogue, asking not only *what* we do and *what* we think but also *why*. As we probe our own traditions, practices, methods, and underlying assumptions, my hope is that we will be empowered to imagine and begin to make positive changes for the future of engineering education.

As a first step, I offer the advice of Ray Kurzweil. With remarkable foresight, he encouraged people in business to take steps to anticipate the future. His advice applies equally to engineers and engineering educators:

> One area of advice that I like to give is to actually take the discipline of writing down what the underlying technologies that affect your business will be a year from now, two years from now, three years from now ... actually take the discipline of writing that out.
>
> The kind of organization that's not going to succeed [is one that is] overly attached to the old business model. ...
>
> Some companies ... have been very successful in redefining themselves continually. You've got to be willing to take those kinds of risks …. (McKinsey and Company [22])

My hope is that this book will provoke us to look optimistically toward the future and at the same time, heed the warning of *Forbes* writer, Ryan Craig:

> in less than a decade we will see the new iteration of ChatGPT writing a nice obituary for decrepit institutions of education that have lost any relevance [9].

May we not be among them.

References

1. Adger, W. N., Fransen, S., Safra de Campos, R., & Clark W. C. (2024, 16 January). Migration and sustainable development. *PNAS,* 121(3). https://doi.org/10.1073/pnas.2206193121
2. Adobe Acrobat Team. (2022, 08 November). Fast-forward—Comparing a 1980s supercomputer to the modern smartphone. *Adobe Blog.* https://blog.adobe.com/en/publish/2022/11/08/fast-forward-comparing-1980s-supercomputer-to-modern-smartphone
3. Agoda Careers. (2024, 30 April). The role of mentorship in engineering. Agoda. https://careersatagoda.com/blog/tech/the-role-of-mentorship-in-engineering/
4. Appleby, C. (2023, 07 March). Will colleges ban ChatGPT? BestColleges. https://www.bestcolleges.com/news/will-colleges-ban-chatgpt/
5. Bruner, J. (1960). The process of education. Harvard University Press.
6. Castillo, Evan. (2023, 27 March). These schools and colleges have banned Chat GPT and similar AI tools. BestColleges. https://www.bestcolleges.com/news/schools-colleges-banned-chat-gpt-similar-ai-tools/#schoolswithdrawn

7. Cloud computing vs. cloud storage. (n.d.). *PureStorage*. https://www.purestorage.com/knowledge/cloud-computing-vs-cloud-storage.html#:~:text=Cloud%20computing%20requires%20a%20large,but%20faster%2C%20more%20efficient%20storage.
8. Copernicus. (2024, 08 February). In 2024, the world experienced the warmest January on record. *Copernicus Climate Change Service*. https://climate.copernicus.eu/copernicus-2024-world-experienced-warmest-january-record
9. Craig, Ryan. (2023, 21 April). Chataclysm: How AI will upend entry level jobs. *Forbes*. https://www.forbes.com/sites/ryancraig/2023/04/21/chataclysm-how-ai-will-upend-entry-level-jobs/?sh=ba06cb41317d
10. Filho, W. L., Salvia, A. L., Beynaghi, A., Fritzen, B., Ulisses, A., Afila, L. V., Shulla, K., Vasconcelos, C. R. P., Moggi, S., Mifsud, M., Anholon, R., Rampasso, I. S., Kozlova, V., Iliško, D., Skouloudis, A., & Nikolaou, I. (2024). Digital transformation and sustainable development in higher education in a post-pandemic world. *International Journal of Sustainable Development & World Ecology*, 31(1), 108–123. https://doi.org/10.1080/13504509.2023.2237933
11. Friedman, A. (2023, 04 December). Intel could create a chip with one trillion transistors by 2030 says its CEO. *PHONEARENA*. https://www.phonearena.com/news/intel-chip-one-trillion-transistor-count-2030_id153786
12. Friedman, T. (2016). *Thank you for being late: An optimist's guide to thriving in the age of accelerations*. Picador.
13. Gilligan, C. (2023, 18 May). How the pandemic boosted working from home. *U.S. News & World Report*. https://www.usnews.com/news/health-news/articles/2023-05-18/how-the-covid-pandemic-impacted-working-from-home
14. Goodall, M. (2020, 20 May). A history of online communities. *Guild*. https://guild.co/blog/history-of-online-communities/#:~:text=The%20forebears%20of%20today's%20online,right%20from%20the%20very%20start!
15. Grassini, S. (2023). Shaping the future of education: Exploring the potential and consequences of AI and ChatGPT in educational settings. *Education Sciences*, 13 (692), pp. 1–13. https://www.mdpi.com/2227-7102/13/7/692
16. Howarth, J. (2024, 02 July). 57 new artificial intelligence statistics (Jul 2024). *Exploding Topics*. https://explodingtopics.com/blog/ai-statistics
17. iPhone 15. (n.d.) Wikipedia. https://en.wikipedia.org/wiki/IPhone_15
18. Jai, B-R., & Shih, M-F. (2024). Technology: Limited or Infinite? *Emerging Media*, 2 (1), pp. 55–69. https://doi.org/10.1177/27523543241246586
19. Jaakkola, H., Henno, J., & Mäkelä, J. (2024, 20–24 May). *Globalization and Education. [Paper presentation]*. 47th MIPRO ICT and Electronics Convention (MIPRO), Opatija, Croatia. https://ieeexplore.ieee.org/abstract/document/10569521
20. Marinkovic, J. (2023, 09 October). The shifting importance of soft skills in a time of AI systems, large language models and global AI regulations. *ISACA*. https://www.isaca.org/resources/news-and-trends/isaca-now-blog/2023/the-shifting-importance-of-soft-skills-in-a-time-of-ai-systems
21. McKenzie, J. (2023). The price of Moore's law. *Physics World*, 36 (8), 30–34. https://doi.org/10.1088/2058-7058/36/08/25
22. McKinsey & Company. (2011, January). IT growth and global change: A conversation with Ray Kurzweil. *McKinsey Quarterly*. https://www.mckinsey.com/~/media/McKinsey/Business%20Functions/McKinsey%20Digital/Our%20Insights/IT%20growth%20and%20global%20change%20A%20conversation%20with%20Ray%20Kurzweil/IT%20growth%20and%20global%20change%20A%20conversation%20with%20Ray%20Kurzweil
23. Mortensen, O. (2024, March). How many users does ChatGPT have? Statistics & facts (2024). *SEO.AI*. https://seo.ai/blog/how-many-users-does-chatgpt-have#:~:text=The%20most%20recent%20data%20from,users%20on%20a%20weekly%20basis.

24. Nicolle, W., Werden, L. & Hayhoe, K., (2024, 09 January). Bloom or bust. UBS. https://www.ubs.com/global/en/sustainability-impact/sustainability-insights/bloom-or-bust.html
25. Pursnani, V., Sermet, Y., Kurt, M., Demir, I. (2023, 08 November). Performance of ChatGPT on the US fundamentals of engineering exam: Comprehensive assessment of proficiency and potential implications for professional environmental engineering practice. *Computers and Education: Artificial Intelligence,* 5. https://www.sciencedirect.com/science/article/pii/S2666920X23000620?via%3Dihub
26. Pushkar, A. (2024, 15 May). Top 30 artificial intelligence (AI) tools list. *IntelliPaat*. https://intellipaat.com/blog/top-artificial-intelligence-tools/
27. Reid, J. (2024, 10 July). Number of millionaires to soar globally but plunge in the UK, research finds. *CNBC*. https://www.cnbc.com/2024/07/10/wealth-report-millionaires-to-soar-globally-but-plunge-in-the-uk.html
28. Smite, D., Moe, N. B., Jildrum, J., Gonzalez-Huerta, J., Mendez, D. (2023, January). Work-from-home is here to stay: Call for flexibility in post-pandemic work policies. *Journal of Systems and Software,* 195. https://www.sciencedirect.com/science/article/pii/S016412122200228X/pdfft?md5=b6ed84b4c76cae9e6b12756f8886b08b&pid=1-s2.0-S016412122200228X-main.pdf
29. The evolution of artificial intelligence. (2024, 30 January). *WeAreTechWomen*. https://wearetechwomen.com/the-evolution-of-artificial-intelligence-from-origins-to-2024/
30. The Institute for Teaching and Learning. (2024). The Socratic Method: Fostering critical thinking. *Colorado State University*. https://tilt.colostate.edu/the-socratic-method/#:~:text=The%20Socratic%20Method%20involves%20a,discussion%20goes%20back%20and%20forth
31. UN biodiversity conference 2024 to feature first-ever "Trade Day." (2024, 23 May). UN Trade & Development (UNCTAD). https://unctad.org/news/un-biodiversity-conference-2024-feature-first-ever-trade-day
32. United Nations. (2024, 11 July). Growing or shrinking? What the latest trends tell us about the world's population. UN News. https://news.un.org/en/story/2024/07/1151971#:~:text=The%20global%20population%20reached%20nearly%208.2%20billion%20by%20mid%2D2024,than%20expected%20a%20decade%20ago
33. Viña, A., & Liu, J. (2022, 23 August). Effects of global shocks on the evolution of an interconnected world. *Springer*. https://link.springer.com/content/pdf/https://doi.org/10.1007/s13280-022-01778-0.pdf
34. Which schools offer free online engineering courses? (2023–2024). Learn.org. https://learn.org/articles/Which_Schools_Offer_Free_Online_Engineering_Courses.html

The manufacturer's authorised representative in the EU is Springer Nature Customer Service Centre GmbH, Europaplatz 3, 69115 Heidelberg, Germany. If you have any concerns regarding our products, please contact ProductSafety@springernature.com

Printed and bound by CPI Group (UK) Ltd, Croydon, CR0 4YY

26/03/2026

02078939-0018